BOOKS ON REAL-TIME SYS
AND DATA TRANSMISSION

by JAMES MARTIN

D1143212

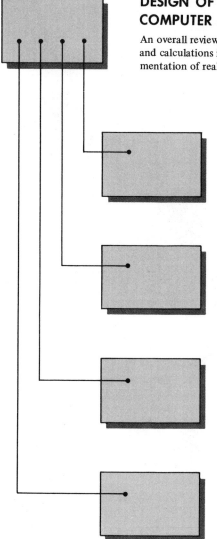

DESIGN OF REAL-TIME COMPUTER SYSTEMS

An overall review of technical considerations and calculations in the design and implementation of real-time systems.

PROGRAMMING REAL-TIME COMPUTER SYSTEMS

Programming mechanisms, program testing tools and techniques, problems encountered, implementation considerations, project management.

TELECOMMUNICATIONS AND THE COMPUTER

A description of the working of the world's telecommunication links and their uses for data transmission.

TELEPROCESSING NETWORK ORGANIZATION

An explanation of the many types of devices and procedures for controlling and organizing the flow of data on today's telecommunication lines.

FUTURE DEVELOPMENTS IN TELECOMMUNICATIONS

An exploration of the foreseeable future in a technology that has reached a period of very rapid change.

THE COMPUTERIZED SOCIETY

(co-author Adrian R. D. Norman)
Euphoria, Alarm, Protective Action. An appraisal of the impact of computers on society over the next fifteen years.

Prentice-Hall
Series in Automatic Computation
George Forsythe, editor

PRENTICE-HALL INTERNATIONAL, INC., *London*
PRENTICE-HALL OF AUSTRALIA, PTY., LTD., *Sydney*
PRENTICE-HALL OF CANADA, LTD., *Toronto*
PRENTICE-HALL OF INDIA PRIVATE LIMITED, *New Delhi*
PRENTICE-HALL OF JAPAN, INC., *Tokyo*

FUTURE DEVELOPMENTS
IN
TELECOMMUNICATIONS

FUTURE
DEVELOPMENTS IN

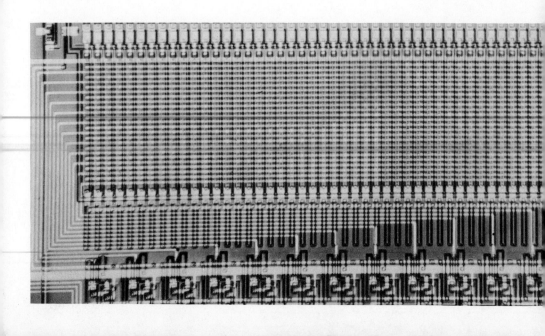

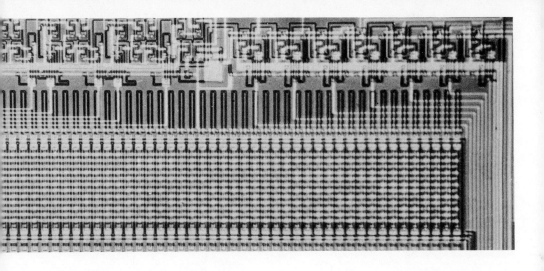

TELECOMMUNICATIONS

by JAMES MARTIN

PRENTICE-HALL INC., ENGLEWOOD CLIFFS, N.J.

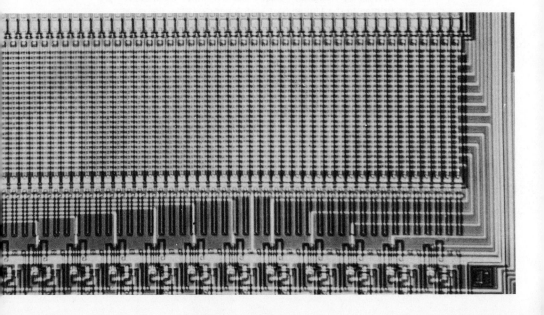

Future Developments in Telecommunications by James Martin

© 1971 by
PRENTICE-HALL, INC.
Englewood Cliffs, N.J.

Current printing (last digit):

10 9 8 7 6 5 4 3 2 1

13–345868–7

Library of Congress Catalog Card Number 74–156757
Printed in the United States of America

TO CHARITY

ACKNOWLEDGMENTS

Any book about the state of the art in a complex technology draws material from a vast number of sources. While many of these are referenced in the text, it is impossible to include all of the pioneering projects that have contributed to the new uses of telecommunications. To the many systems engineers who contributed to this body of knowledge, the author is indebted.

The author is very grateful for the time spent reviewing and criticizing the manuscript by Mr. J. W. Greenwood and Miss C. Howland Anders in New York, Lord Snow and his staff in London, Mr. J. C. McPherson and Mr. W. Short at Armonk, Messrs. L. Hochgraf and J. J. Mahoney of the Bell Telephone Laboratories, Mr. P. Musgrave, and his colleagues at the Datran Corporation, Mr. H. Poppell of Booz Allen and Hamilton, and Mr. Eichenberger and his staff in Zurich. The author is also grateful to the organizations that contributed pictures of historic equipment a few that appear in the chapter openings, and which convey a feeling of how fast our world is changing. The long-suffering students of IBM's Systems Research Institute in New York also used the manuscript and made many comments, but opinions expressed in the book are those of the author, and not of the corporation he works for. Mr. R. B. Edwards' staff helped enormously in typing and reproducing the manuscript. The author is particularly grateful to Miss Cora Tangney for her help in this.

Last, and perhaps most important, the author is indebted to Dr. E. S. Kopley, Director of the IBM Systems Research Institute for his constant encouragement. Without the environment that he created, this work would not have been completed.

CONTENTS

x

HOW TO READ THIS BOOK

This book is intended for a nontechnical as well as a technical audience. Thus technical jargon has been avoided where possible but could not be eliminated totally. Most technical concepts are explained as the book progresses. In addition a glossary is included at the end of the book.

The most technical material appears in Section III. The nontechnical reader can omit this section without destroying the continuity of the book. It is recommended, however, that the technical reader, does not leave it out. The nontechnical reader may also skip over some of the sections relating to data transmission.

Persons unfamiliar with telecommunications technology will probably grasp the concepts of this book more completely if they read the author's *Telecommunications and the Computer* (Prentice-Hall, 1969) in conjunction with this book.

SECTION I

INTRODUCTION

Telecommunications in New York, 1838. This semaphore station on Staten Island was an intermediate between Sandy Hook and Manhattan, seen in the distance.

1 THE COMMUNICATIONS REVOLUTION

Communication technology has entered a period of revolutionary change.

The last decade has brought new inventions of enormous potential. It will probably be two decades before we fully grasp the shattering effect they will have on society.

These inventions include:

The communication satellite. Suddenly this has provided telephone and television links to the underdeveloped world. Much larger satellites will be built and will have an enormous impact on education and communications both in the United States and throughout the world. The satellite antennae in some underdeveloped countries stand next to fields ploughed by oxen.

The helical waveguide. A pipe, now operating, that can carry 250,000 simultaneous telephone calls or equivalent information over long distances.

The laser. This means of transmission, still in the research laboratories, has the potential of carrying many millions of simultaneous telephone calls or their equivalent.

Large-scale integration (LSI). A form of ultraminiaturized computer circuitry that probably marks the beginning of mass production of computers and computerlike logic circuitry. It offers the potential of extremely reliable, extremely small, and, in some of its forms, extremely fast computers. If large-enough quantities can be built, this circuitry can become very low in cost.

On-line real-time computers. Computers capable of responding to

3

many distant terminals on telecommunication lines at a speed geared to human thinking. They have the potential of bringing the power and information of innumerable computers into every office and eventually every home.

Picturephone. A public dial-up telephone system in which subscribers see as well as hear each other.

Large TV screens. TV screens that can occupy a whole wall if necessary.

Cable TV. Provides a cable into homes with a potential signal-carrying capacity more than one thousand times that of the telephone cable. It could be used for signals other than television.

Voice answerback. Computers can now assemble human-voice words and speak them over the telephone. This fact, coupled with the Touchtone telephone set, makes every such telephone a potential computer terminal.

Millimeter-wave radio. Radio at frequencies in the band above the microwave band can relay a quantity of information greater than all the other radio bands combined. Chains of closely spaced antennas will distribute these millimeter-wave signals.

Pulse code modulation. All signals, including telephone, Picturephone, music, facsimile, and television can be converted into digital bit stream and transmitted, along with computer data, over the same digital links. Major advantages accrue from this.

Computerized switching. Computerized telephone exchanges are coming into operation, and computer-like logic can be employed for switching and "concentrating" all types of signals.

Data banks. Electronic storage for huge quantities of information that can be manipulated and indexed by computers and that can be accessed in a fraction of a second.

Few technologies could have as profound an effect on the human condition as the full development of these thirteen inventions, and certainly additional inventions in telecommunications are yet to come, some perhaps of even greater impact. The last three decades of the twentieth century will be remembered as the era when man acquired videocommunication channels to other men, to libraries of film and data, and to the prodigious machines.

The new means of communication have potential both for great good to men, and for great evil. They will forge links between people, raise productivity, and make the best of man's culture available to all. On the other hand, they will also make the worst of man's culture available to all. They provide new techniques of tyranny. They make possible Orwell's telescreen and at the same time make it naive, for Orwell did envision computers.

In the industrialized nations, these devices will provide the means of education needed to keep up with the ever-increasing rate of change—to combat the "future-shock" of too rapid an evolution. In the underdeveloped countries, they will provide the first hope of literacy and afterward a headlong plunge into the technology that made the northern white nations rich. However, along with education and culture, the changes can also bring propaganda, cruelty, and distortion.

International corporations, in the new era, will flourish with the help of worldwide computer networks. Human talent, the scarcest resource in the coming cybernetic age, will be made available worldwide. Communications are essential to a growing world economy. Business and societal patterns will change as screen-to-screen communication proves more efficient than traveling. In the future, traveling will be more for pleasure than for business. Certain types of people may work at home much of the time, dialing computers and participating in teleconferences. In addition, they may shop at home, receive education at home, be entertained at home, and interlink each others' homes with Picturephone. Money transactions will generally be handled without cash or check, and man's credit rating will mean more than his pedigree. Furthermore, our concepts of privacy may change in a world of interconnecting computers.

The satellites and screens will undoubtedly open windows between nations, and wise use of them will do much to forge links and understanding between men. The result may well be worldwide diffusion of knowledge, skills, and culture. Perhaps the best form of foreign aid rich nations can give to poor ones is mass education and training in agricultural and technological skills. The new communication links make this help possible. Five thousand villages in India—many without schools, many without radio or telephone, many without light from sunset to sunrise, and isolated from the world by the inability of their people to read—will soon have a village television set receiving educational programs from a NASA satellite. But television can also increase unrest; it can spread and often amplify chaotic situations. To what extent is the unrest in the United States and Japan amplified by television? What will be the effect on multimillions of previously isolated people in the underdeveloped nations of seeing American affluence and worldwide riots on their screens?

There is little doubt that in the United States the new communication technology will enrich society enormously. It will have a great effect on education; it will increase industrial productivity and make possible many new ventures in industry. It seems likely that in the United States and Canada, and perhaps Japan, the technology described here will be far ahead of that in Europe and elsewhere. This fact, along with computers, will further enhance the competitive edge of the technically advanced nations. The investment in new forms of education alone will produce high dividends.

As industrial automation grows, a higher proportion of people will work in service industries, and there will be an ever-increasing amount of leisure. Many service industries will be dependent on or based on the techniques discussed in the chapters ahead. Education for leisure will be vitally important, for it will be a moribund society where leisure becomes equated with passive television watching. The communication links will be vital for education and computer-dependent leisure activities, and the media in the home must be interactive rather than passive.

To the accepted stages of economic growth of a society, a new one will be added in time. Societies will modify their behavior patterns when a state of "saturation of affluence" is reached. Eventually, as the Gross National Product per capita rises, through automation, the mass striving for more money and material goods must give way to other drives. A bigger electric can opener does not give life a richer texture. Music, opera, education in the arts and literature, informed conversation, leisure facilities to explore an intellectual world, and entertainment and hobbies using the new media, do. The Protestant ethic will have given way, perhaps to attitudes closer to those of Athens in its prime, perhaps to new attitudes unclassifiable in terms of past history. Man will have the opportunity to once again become civilized.

To provide the new services, new telecommunication links are needed. When the links are there, the services will grow. The greater the usage of the links, the lower their cost, for the new technology brings great economies of scale. On the other hand, some services will not be initiated *until* the costs of the links are low. To some extent we have a chicken and egg argument. Which comes first, the telecommunication network or the services using it?

Physical communications like railways and canals had a major effect on the growth of the Industrial Revolution. To a large extent, they made it possible. The same is true in the electronic revolution, and unavailability of communication links will impede progress. In my work in the design of computer systems, I have seen many splendid schemes fall by the wayside because the telecommunication links needed were not available or were too expensive. As will be illustrated, incomparably better data transmission facilities than those now available could be provided, and the result would have a major effect on industry.

Recently I returned from a developing island that as yet has no doctor and almost no roads. While there, I heard a conversation between a sage local black leader of the community and a benevolent American on the subject of which should come first: a doctor or a road across the island. The American insisted that a doctor was all important, and the local man said there was no point in having a doctor until the road existed so that he could get to his patients. The latter view will prevail. The island cannot develop until it has its road.

Lack of adequate telecommunication links or suitably priced tariffs suppress the development of applications and their marketing. Such links will have a major effect on productivity, economic growth, and the lives of people. They are the catalyst of the electronic revolution. There are few better investments that an advanced country could make at this point in time than extensive spending on a dial-up data transmission network and other such facilities described in the chapters ahead.

"Mr. Watson, come here; I want you"—the first articulate sentence ever spoken over an electric telephone by Alexander Graham Bell, March 10, 1876, spoken after he spilled on his clothes some of the acid that was part of this transmission apparatus. Courtesy Western Electric.

2 THE CRYSTAL BALL

The intent of this book is to take all the facts that seem relevant to telecommunications, explain them to the reader, and piece by piece build a picture of where we are going and what is likely to be achieved. Before embarking on this it would be as well to see what success others have had at predicting the directions of technology, and establish some ground rules.

There have been some spectacular failures.

When Thomas Edison announced that he was working on an incandescent lamp, gas securities dived and the British Parliament set up a committee to investigate. The committee reported that Edison's ideas were "good enough for our trans-Atlantic friends . . . but unworthy of the attention of practical or scientific men." The chief engineer of the British Post Office called it *ignis fatuus*.

The airplane, the telephone, the space rocket, the atomic bomb, the computer, and radio and television broadcasting all met similar derision when they were discussed shortly before becoming practical realities. The derision usually came from leading scientists or engineers of the day and often from committees of them.

In 1956 the British Astronomer Royal Dr. Woolley announced to the press that "Space travel is utter bilge" [1]. This remark was made *after* President Eisenhower had announced the United States satellite program. The very next year the Russians launched Sputnik I. Lack of forecasting accuracy does not seem to matter; Dr. Woolley later became a leading member of the committee advising the British government on space.

9

Although several thousand German V2 rockets had blitzed London and Belgium in World War II, causing great loss of life, Dr. Vannevar Bush advised the United States Senate in December 1945 that a 3000-mile-range bomb-carrying rocket was "impossible." He said, "I say, technically, I don't think anyone in the world knows how to do such a thing . . . I think we can leave it out of our thinking. I wish the American public would leave it out of their thinking."

Winston Churchill's scientific adviser, Professor F. A. Lindemann (Lord Cherwell), had given the House of Lords the same misinformation months earlier. He had not learned from the fact that nine months before the V2s rained down on London, he had advised Churchill that he "discounted the probability of the use of large rockets" [3]. He refused to believe that the German V2 could fly, although there was photographic evidence that it could.

FAILURES
OF NERVE

Arthur C. Clarke, the science-fiction author, divides forecasting failures into two classes: "failures of nerve" and "failures of imagination" [1]. The former case occurs, he says, when, *given all of the relevant facts* the would-be prophet cannot see that they point to an inescapable conclusion. He often refuses to believe that anything fundamentally new can happen. "Failures of nerve" are frequently accompanied by substantial emotion. The more spectacular illustrations, such as the preceding ones, often come from older persons who are more reluctant to change preconceived ideas.

Robert U. Ayres points out in his book *Technological Forecasting* [4] that committees for forecasting are often prone to failure of nerve. The first major forecasting effort by committee was conducted in 1937 by the U.S. National Research Council [5]. It produced a sober and responsible document, which began "In an age of great change, anticipation of what will probably happen is a necessity for the executives at the helm of the Ship of State" and which missed virtually every major development of its decade, including antibiotics and radar (both of which had existed for ten years), jet engines (which had been designed in theory), and atomic energy (which had been much speculated about). In 1940 a National Academy of Sciences Committee was set up in the United States to evaluate the proposed gas turbine [6]. The committee concluded that the turbine was quite impractical because it would have to weigh 15 lb/hp (pounds per horsepower) as opposed to 1.1 lb/hp for internal combustion engines. One year later a gas turbine was in operation in England

and it weighed 0.4 lb/hp. One finds corporate strategy reports today that are almost as erroneous.

The presumed advantage of a committee is that interactions between the members will produce a creative synergism and that individual biases and hobbyhorses will be averaged out. In fact, however, overconservatism usually seems to result. In a ten-year study of economic forecasting, the U. S. National Bureau of Economic Research concluded that individuals do consistently better than consensus polls, at least on economic issues [7]. Ayres comments "All revolutionaries are aware of the Leninist warning that the most dangerous threat to its leadership always comes from the radical left. On technical committees, the unspoken rule is generally to avoid being outflanked on the right (i.e., by the more conservative elements)!"

We are all, in differing degrees, prisoners of familiarity. This fact is often true of the technologist viewing his own discipline and the industrialist viewing his own product. The technologist and industrialist understand in such fine detail how things are done today that they cannot imagine the sweeping changes that will come tomorrow; the detail of these is not yet known. They focus on the limitations set by the current state of the art.

FAILURES OF IMAGINATION Arthur Clarke's second category, failures of imagination, applies with a vengeance to the data processing and communication industries. When the computer came into existence at the end of the 1940s, confounding all the forecasts that said it was a technical impossibility, very few people had the imagination to see how it could be used.

Forecasters estimated a commercial market of not more than 12 machines in the United States. "Only ten or a dozen very large corporations will be able to take profitable advantage of the computer" was a view expressed in 1948. Sometime later IBM (International Business Machines) made an historic decision not to market the computer because it would never be profitable. The problem was that people failed to see how the machine would be used; they lacked the imagination to think of suitable applications. They thought of the machine as doing only scientific calculations and could only visualize a small number of calculations that were big enough. As the computer industry has grown, forecast after forecast has suffered from a similar, if less spectacular, failure of imagination. When disk memory had been available for a few years, a leading computer salesman in England assured me that it was an

American gimmick that would soon disappear from the marketplace. (This person now has a high management position.) When real-time systems were working in airlines and savings banks, it was a commonly held view that there would be no application for real-time methods other than in airlines and savings banks. Similarly, when data transmission first became available, one heard repeatedly that nobody in his right mind would use it because the mail was cheaper.

The same failure to foresee applications has plagued the industries concerned with transmitting information. When broadcasting was first proposed earlier this century, a man who was later to become one of the most distinguished leaders of the industry announced that it was very difficult to see uses for public broadcasting. About the only regular use he could think of was the broadcasting of Sunday sermons, because that is the only occasion when one man regularly addresses a mass public. At many times in telephone history, engineers have considered their latest high-capacity channel the ultimate, and have failed to imagine how the public could use greater capacity. Consequently, new applications have often failed to materialize because of bandwidth shortage. One still sees forecasts for data transmission that apparently regard it as an extension of telegraphy, and fail to realize that data transmission will have fundamentally new uses. The corporate strategy reports in the information industries today seem virtually guaranteed to be failures of imagination.

TECHNOLOGICAL SURPRISES   In the preceding categories, we discussed failures to forecast when all the facts are known.

In addition, technology has surprises in store for us that are likely to be entirely unpredictable. Some of these surprises contradict the most cherished views of their time. In this century several laws that were regarded as the most fundamental foundation stones of physics have been proven wrong. My dictionary still defines an atom as "a particle of matter so minute as to admit of no division." At school I was taught "matter can neither be created nor destroyed." The Heisenberg uncertainty principle, the general theory of relativity, quantum mechanics, the discovery of mesons, the quasar, and perhaps the pulsar, have all cracked the foundation stones of their day. Two years ago I would have argued vehemently that nothing could travel faster than light. Now the science journals are full of discussion of faster-than-light particles.

However, a technological surprise does not need to crack the foundation stones in order to play havoc with the best-laid plans of mice, men, and AT & T. Although he might possibly have realized that they were conceivable in theory, a reasonable forecaster in 1940 would not have predicted the computer; in 1945 he would not have predicted the transistor; in 1950 he would not have predicted the laser; in 1955 he would not have predicted the use of pulse code modulation, large-scale integration, solid state switching, computer time sharing, on-line real-time systems in commerce, direct-access data banks or synchronous communication satellites; in 1960 he would not have predicted holography. All of these inventions have a major effect on the story we have to tell.

Today's computers would have been quite inconceivable in 1940. The idea would have been laughed at as the wildest fantasy. Yet within two decades the data processing industry in the United States is likely to exceed the Gross National Product of 1940.

The rate of developing new technology is increasing constantly. We can expect the number of technological surprises in the two decades ahead to be greater than the number in the past two decades.

UNDERESTIMATING DEVELOPMENT TIME   On the other side of this argument, an important lesson to learn from technological history is that the appearance of a surprise invention does not immediately make existing equipment obsolete. In fact, there is quite a lengthy time between the conception of a complex invention and its use in public systems. The first working laser was developed in 1959. Devices using it were being sold in the second half of the 1960s, but public telecommunication channels based on it are unlikely to be working before the early 1980s.

In the provision of highly complex and expensive systems, such as new communication channels, we can expect a time lag of some years because of three factors.

1. Several years of refining a basic invention are needed before it is suitable for such systems, and several years more are needed to perfect manufacturing processes and bring down costs. With many new technologies—with the laser in fact—more than a decade of research is needed before it becomes a suitable basis for system development.
2. Very large blocks of capital are needed for the equipment and its manufacturing processes. There is a reluctance to commit these until the economics of the new techniques are proven.
3. Particularly important in telecommunications, there is a vast investment in present-day plant, which cannot simply be discarded when

a new technology appears. A problem today is that one technology is following another at an increasingly rapid rate. Telecommunication plant has traditionally been designed to have a 40-year lifetime. Fortunately, the growth rate is sufficiently fast that this lifetime does not prevent the introduction of new types of system, but it does mean that the old equipment must co-exist with the new.

The development lead time (the time taken from the start of development to the first commercial application) is sufficiently great that surprise inventions coming this year would almost certainly not find their way into public communication channels and switching during the next decade.

UNDERESTIMATING THE COMPLEXITY   If excessive conservatism and failures of nerve prevent predictions of change, on the other hand, once a new concept is grasped would-be forecasters often let their enthusiasm run away with them. Not only is the development lead-time forgotten, but there is also an almost inevitable tendency to underestimate the complexity of implementation. The press and popular books about the future are full of this kind of oversimplification. The gee-whiz school of forecasting suffers from a failure to work out the details. A simple-seeming end product such as Picturephone or the ability to dial up movies for showing on the home television screen can be grasped without appreciation of the detailed difficulties that are entailed.

One can look back ten years at the forecasts then made for computers. Today's applications are much more diverse than these predictions (failure of imagination). Today's hardware is faster, more reliable, cheaper, and has more on-line storage than was forecast. The machines make a major unforecast use of telecommunication links and terminals. However, in one aspect they entirely failed to live up to the forecasts—that is, in the imitation of intelligent human functions. We have not come close to the predictions made for language translation, speech recognition, associative memories, and other forms of "artificial intelligence." The gee-whiz forecasts for computers writing Western scripts have not been fulfilled. The reason is that these are all functions that we can imagine easily because they are human functions, but we totally underestimated how complex they really are. This complexity was painfully discovered during the years of trying to program them.

We must make a cautious assessment of the complexity of schemes eligible for discussion in this book.

LEGAL AND   Many worthy schemes have foundered on po-
POLITICAL   litical and legal rocks, and the waters we navi-
PROBLEMS   gate in discussing communications are filled
with such rocks.

Multibillion dollar vested interests are at work. The American Telephone and Telegraph Company has no intention of losing a vast revenue to satellites. The television networks see grave dangers from cable TV. The computer companies live in fear of the Federal Communications Commission (FCC) or other government regulation. The term "computer utilities" is taboo in some circles. The heavily regulated common carriers have no intention of allowing small companies to move in and skim the cream of the business using new equipment.

Some of these factors will be discussed later, but generally it is so uncertain who will win the battle of politics that most of the book concentrates on what is *technically* possible. As will be discussed in the last chapter, it seems sadly probable that although great strides will be made, the development of this vital technology will be far from optimal because of political scuffling.

FAILURE TO   Another failure in forecasting is caused by
FORECAST MARKET   market constraints. Sometimes a viable and
CONSTRAINTS   interesting product is never placed on the
market. This may be due to lack of entrepreneurial interest. It may be because it cuts across a large corporation's existing market, or because the profit margin has been judged not high enough.

Sometimes inventions related to telecommunications can only be marketed by a large common carrier like AT & T. This fact would be true, for example, for innovations in telephone switching discussed in Chapter 19. There have been many ideas for improving the service of the telephone network which have never reached the marketplace. It would be useful, for instance, and not difficult technically, to have automatic message recording at the local central office (telephone exchange). If a subscriber were out when called, his caller could leave a spoken message, which would be automatically recorded, and that subscriber could retrieve it when he returned, for a small fee. Although AT & T did not implement this relatively simple scheme, they did implement the Picturephone® service, which was exceedingly difficult technically, very ex-

®Registered Trade Mark.

pensive, and some say, of dubious market potential. You never can tell.

Similarly, some of the ideas discussed in this book may never reach the marketplace; on the other hand, the majority of them probably will. One could certainly invent many ingenious gadgets that use telecommunications but that would be unlikely to be marketed. I would like to have a grand piano, for example, capable of being played either by a pianist or by a stream of digital impulses coming over a telephone line— a teleprocessing piano. The user could dial a computer and request Rubinstein playing Mozart on his home piano, or he might play a duet with the computer when he is learning, the machine taking the left hand, for example. He could prerecord certain trills or phrases and make the machine play these by pressing a single key. He could listen on his own piano to distant friends playing or could play duets with them. The teleprocessing piano has many possibilities. The computer-minded reader might like to work out how he would organize the bit stream and buffer, assuming no more than 3600 bits per second on the telephone line.

It is virtually certain that the teleprocessing piano will not come to the marketplace, however, because too few people would be prepared to pay for it. When other possible services are discussed in the chapters ahead, reader should ask himself: Is this something that can be sold, or is it just another teleprocessing piano?

LONG-TERM TRENDS   In many complex technologies, a long-term trend can be plotted as a relatively smooth curve in spite of "surprises" that occurred on the way. Often a growth rate in capability close to an exponential curve can be traced.

If we can take *one class* of invention, a measure of its performance often follows a curve like that of Fig. 2.1. The speed of the automobile and the power of vacuum-tube computers followed such a curve, for example. After a slow start, technical improvements gather momentum until saturation eventually sets in and it is either technically or economically unfeasible to increase the performance of the invention further.

In many technologies, however, new classes of invention replace earlier ones for performing the same function. In looking at the performance of one class by itself, we find that it follows a curve like that in Fig. 2.1, but in studying the change as a whole, we must consider the effect of many such curves superimposed as each class of invention replaces the previous one. This situation is illustrated in Fig. 2.2, where four generations of technology or classes of invention replace one another. Each, by itself, is

tion IV looks at the more distant future. Chapter 22 discusses the
tial of the next 30 years and some of this potential must include
iques entirely beyond the knowledge of today's engineers.

does, however, seem reasonable to make predictions based on
polation of long-term trends, realizing that these can only be broad,
etailed statements. Because our legal, political, and societal struc-
is extremely slow to change, it is of value to project the technology
decades into the future if we can. Only by looking well ahead can
ope to direct our turbulent evolution into channels that will make
s world a better place to live in.

e can make one firm prediction. The change in communications
g this time span will have a major effect on the life of every indi-
al.

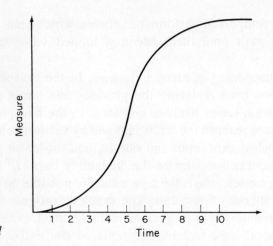

Figure 2.1

REFERENCES

Arthur C. Clarke, *Profiles of the Future* (London: Victor Gollancz Ltd.,
1962).

Vannevar Bush, testimony before the Special Committee on Atomic
Energy, December 1945.

Winston S. Churchill, *The Second World War. Closing the Ring* (Boston:
Houghton Mifflin, 1951), p. 239.

Robert U. Ayres, *Technological Forecasting* (New York: McGraw-Hill,
1969).

W. C. Ogburn et al., "Technological Trends and Nation Policy," U.S.
National Research Council, Natural Resources Committee, 1937.

Technical Bulletin No. 2. U.S. Navy, Bureau of Ships, January 1941.

Victor Zarnowitz "An Appraisal of Some Aggregative Short-term Forecasts," National Bureau of Economic Research, December 1964.

limited. Developments in engineering carry its performance so far, but
then the natural restrictions of the technique place a limit on further
improvement of performance. Meanwhile, however, a new technology
comes into being. The performance curve for the industry as a whole is

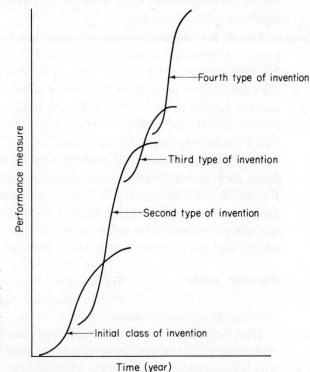

Figure 2.2. As one type of
invention or technology replaces another the growth in
performance continues for a
sustained period. This happened in computers, for example, as one "generation"
has replaced another.

the envelope of the component performance curves, which tends to maintain exponential growth until there becomes limited value in further development.

This process is happening in many industries. In the computer field, one "generation" has been replacing the previous one about every six years, and the situation seems likely to continue. In the first generation, price and performance reached the limits imposed by vacuum tubes, delay line memories, punched card input and output, and bit-by-bit programming. In the second, transistors broke the "reliability barrier," magnetic cores gave fast memories, magnetic tape made it possible to feed the machine as fast as it could digest data, and symbolic programming languages raised the productivity of programmers. In the third, price dropped another tenfold, "solid logic technology" replaced transistors, machine organization raised throughput and allowed attention to many jobs at once, and higher-level languages brought the machine direct to the user. Simultaneously, communications between men and machines have greatly improved. Such a pattern of development will almost certainly continue. We now look forward to large-scale integration, solid-state files, a high degree of parallelism, data management systems, and other new ideas that change the technology.

Figure 2.3 illustrates a similar sequence of developments in nuclear particle accelerators. Figure 22.1 (page 342) shows that such a sequence has been going on for a long time in telecommunications. In this way, the capability of computers, the energy achieved by nuclear accelerators, and the capacity of telecommunication highways have each followed an overall growth curve close to exponential. Many other technologies have had a similar exponential growth rate.

Although we cannot anticipate the surprise inventions, it does, in some cases, seem reasonable to extrapolate the exponential performance curves. Figure 22.1, for example, has been approximately a straight line for so long (the vertical scale being plotted logarithmically) that it seems reasonable to assume that it will continue to grow in this way for some time ahead, and basic research now taking place yields credence to this view.

GROUND RULES   We are now in a position to lay down some ground rules for our discussion of future developments in telecommunications.

First, because of the development lead time in this field, we can expect that all types of channels in service in the next ten years will be based on a technology that is now known. They will be engineered differently than

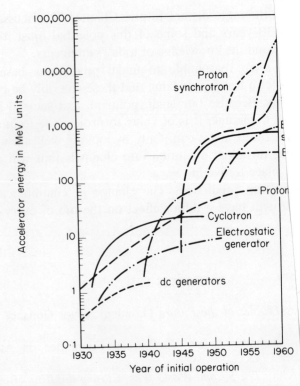

Figure 2.3.   The rate of increase of operating ene ticle accelerators. *From M. S. Livingston.*

today in some cases, but the potential improvements in not invoke major "surprises." On the other hand, such pear at an increasing rate, and thus the types of chan available in twenty years time and beyond will probably tionary new technology. The technology of 1990 is as un (1970) as the transistorized computer with mass on-line was before World War II.

Therefore we have subdivided this book so that Sect the facilities likely to be available during the next ten yea section we may be safe from major surprises, although new uses of communications are planned for the next deca be safe from "failures of imagination" and can be certain nels will be used in ways we have not yet thought of.

Section III discusses the known technology in more d some aspects of it which may not be in use in the next ten y

THE NEAR FUTURE

In 1897, Marconi succeeded in communicating by wireless eight miles across the Bristol channel with this apparatus—the first wireless transmission across water. The upper two pieces are a Righi spark gap (left) and an induction coil. Below is a Morse key. Courtesy The Marconi Co., Ltd.

3 NEW USES FOR TELECOMMUNICATIONS

In the years ahead we will have telecommunication channels of enormous capacity and logic circuitry of high complexity capable of manipulating what happens on them. These channels will change the way men communicate. We no longer think of telecommunications as meaning solely telephone and telegraph facilities. Let us begin the exploration by listing the different forms of communication that are now possible.

Table 3.1 gives the major uses for telecommunication links.

Some of these need the ability to *switch* connections between subscribers, as on the public telephone network. Others do not use a switched circuit: they may use a private or leased channel; they may pick up broadcast signals as the radio does; or they may be connected to a multiuser channel, as with cable TV. Some telecommunication users require two-way channels like the telephone, and others passively pick up one-way signals, as in the case of broadcast television.

Again, different types of signals need different channel capacities. These are normally described in terms of the required *bandwidth* (range of frequencies). If an AM radio reproduces sound frequencies between 100 and 5000 hertz (cycles per second)[1], we would say that it has a bandwidth of 4900 Hz. On the other hand, as any hi-fi enthusiast knows, to reproduce music perfectly one needs to transmit frequencies from about

---

[1]Today the term "cycles per second" is no longer used. 1 hertz (Hz) means the same as 1 cycle per second.
1 kilohertz (kHz) is one thousand cycles per second (kilocycles)
1 megahertz (MHz) is one million cycles per second (megacycles)
1 gigahertz (GHz) is one billion cycles per second.

Table 3.1   MAJOR USES FOR TELECOMMUNICATIONS LINKS

Service	Switched?	One Way or Two Way	Bandwidth (kilohertz)
Telephone	Yes	2	4
Sound broadcasting (low fidelity)	No	1	5
Sound broadcasting (high fidelity, stereophonic)	No	1	40
Music library	Possibly	1	40
Television broadcasting (color)	No	1	4600
Large wall screen television	No	1	20,000 to 50,000
Closed circuit television intercommunications	Possibly	1 or 2	4600
Movie selection in the home	Possibly	1	
Stored television	Possibly	1	1000 to 4600
Still pictures on television screen	Possibly	1 or 2	4 to 100
Facsimile	Possibly	1 or 2	4 to 1000
Teaching devices	Yes or No	2	Any, commonly 4
Advertising	Yes or No	1	Any
Shopping	Yes	1 or 2	4 to 4600
Mobile communications	Yes or No	2	25
Voting by the public	Yes or No	2	<1
Alarms (fire, burglar, system failure, etc.)	Yes or No	1	<1
Emergency communications	Yes	1 or 2	4
Telegraphy	Yes or No	2	<1
Batch data transmission	Yes or No	2	<1 to 1000
Access to time-shared computers	Yes or No	2	<1 to 4
Real-time systems, such as airline reservations	Yes or No	2	Commonly 4
Fast alphanumeric man-computer "conversation"	Yes or No	2	4
Man-computer conversation with graphics	Yes or No	2	4 to 1000
Man-computer conversation with voice answerback	Yes or No	2	4
Banking and credit systems	Yes or No	2	<1
Data collection systems	Yes or No	1 or 2	<1
Intercommunication between computers	Yes or No	2	Any
Automatic meter reading (utilities)	Yes	1	<1
Vehicle traffic control	No	2	<1
(or perhaps alternatively:)	No	1	1000 to 5000

30 to 20,000 Hz (even if the top frequencies are only heard by passing bats!), which is a bandwidth of about 20 kHz. Color television needs a much higher bandwidth—about 4.6 MHz—whereas the telephone has a

low bandwidth, somewhat less than 4 kHz. The bandwidths needed for different types of signals are given in Table 3.1.

Certain items in the table will be discussed later on. The categories "music library" and "movie selection in the home" refer to a system that enables one to *request* particular pieces of music or movies and have them transmitted to one's home, probably over a cable. As we shall see later, this step seems economically feasible today with music but is not likely to be so with movies for at least a decade because movies require a much higher bandwidth. However, movies might be transmitted over a television channel at off-peak hours and recorded automatically for later playback, a process referred to as *stored television*.

MOVIES IN CARTRIDGES   The widespread use of phonograph records and tape cartridges removes much of the requirement to send music to the home over telephone lines. Only now is the on-line music library beginning to appear to be an attractive alternative to having one's own record collection.

A similar argument applies to movies. A mass market will develop in the 1970's for movies recorded on a medium that enables them to be played back over a television set. Officials of RCA expect this movie cartridge industry to reach $1 billion by 1980, and other forecasters put the figure higher. The first such scheme used Columbia Broadcasting System's EVR cartridge. Miniaturized film with a playing time of 25 minutes is distributed in sealed cartridges. The same cartridge could contain several hundred books filmed one page per frame. Four other different and incompatible technologies are competing for the same market: super 8 mm film, magnetic video tape, plastic video disks, and RCA's SelectaVision which records on vinyl tape with laser beams and holography.

Although cartridge movies may dominate the consumer movie market in the 1970s, the sending of movies by telecommunications to individual subscribers may be the ultimate means of distribution when a suitable network exists. Schools in certain areas will be using cables in the 1970's to access filmed educational matter. The same may be true in industry. The cable TV channels being built today may become the means by which the consumer can request his own movies, but this seems unlikely in the next decade.

SLOW SCAN   Stored television could conceivably be used in order to employ a transmission medium slower than a television channel. A dial-up Picturephone channel (Chapter 4) might be used, for example, after midnight to transmit a movie for television viewing. The movie would not be viewed over the small Picturephone screen because the quality would not be as good as over television. Instead, it would be videotaped for later playback. The Picturephone line has only one sixth of the bandwidth needed for television, and thus the movie would be "slow scanned." "Slow scan" is the sending of a motion picture at a slower rate than that at which it will be viewed. This is done in order to use a lower bandwidth than when the picture is sent at full speed. A two-hour television movie would take 12 hours to transmit over Picturephone lines. A recording tape could be automatically switched off after the transmission (and perhaps automatically switched on also). Sometimes when television channels are not available, news film sequences have been sent by slow scan over ordinary telephone lines for later broadcasting by the television networks. This has been done from foreign countries having no satellite channel. The film clipping was usually very short because of the great difference in bandwidth between television and telephone. Today planes carrying film, and satellite transmission are normally used.

Slow scanning can be used, if it is worthwhile, on any transmission that is not listened to or viewed *as the transmission takes place*. In this way, the bandwidth requirements are lowered. With Picturephone, it is being used in experimental systems to improve the resolution of the image received. The same transmission facility and bandwidth are employed, but the receiving device has a larger screen (or copier); and instead of transmitting 30 frames per second, one frame per second or less can be used.

FACSIMILE   *Facsimile* refers to the transmission and reproduction of documents. This may be Xerox machines transmitting to other Xerox machines. It may be the transmission of newspaper photographs or engineering drawings. The speed at which the document is reproduced depends on the bandwidth of the transmission facility, how efficiently it is used, and the degree of resolution in the transmitted document. A typical office machine transmits over a telephone line, public or leased. With this bandwidth, the transmission of an 11 inch × 8½ inch document takes about 6 minutes and gives a reproduction quality equivalent to that of a Xerox copier. If the bandwidth were 12 times as wide (an available broadband tariff), the document could be sent in half a minute. Over a Picturephone line, which has a bandwidth 250 times that of a telephone line, the document could be sent in about 1.4 seconds. The transmission line could keep up with today's Xerox copiers.

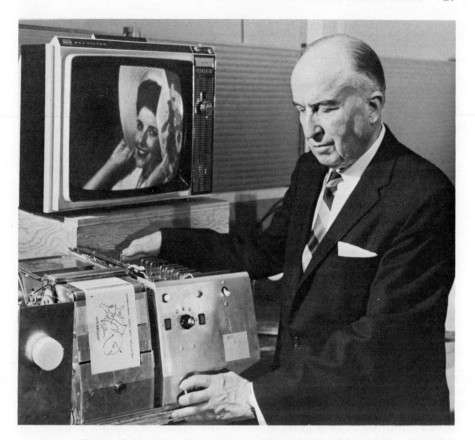

Figure 3.1. RCA Experimental Home Facsimile System. RCA has developed an experimental television system that makes it possible to broadcast printed copy and standard programs to the home simultaneously and over the same channels by means of an electromagnetic "hitchhiking" technique. An electrostatic printer associated with the TV receiver picks off the print signal from the TV antenna without disturbing the TV program and converts it to printed form.

**BANDWIDTH
AND BITS**

It is important to understand these bandwidth calculations. The amount of information transmitted is proportional to the bandwidth multiplied by the transmission time. However, there are many different ways of organizing the information—ways that differ widely in their efficiency. One way, which plays a major role in the story we have to tell, is to convert the signal into a series of bits (one bit is an *on* or *off* pulse—for example, the presence or absence of current) that are transmitted digitally, as in the flow of data in computer circuits. Any of the items in Table 3.1 could be sent this way, as will be discussed in Chapter 6. On the other hand, data that are basically digital in nature, such as that from a computer, could be (and usually are) sent over non-digital telecommunication channels, by appropriate conversion.

Signals needing a large bandwidth need a large number of bits per second to carry them. However the number of bits per second needed is not necessarily proportional to the bandwidth needed for different signal types, as we will explain later.

MULTIPLEXING   A channel of a given capacity can be split up into a number of lesser channels. A given bandwidth or range of frequencies can be split up into smaller bandwidths by a technique called *frequency-division multiplexing* [1]. Similarly a bit stream can be subdivided to carry lesser bit streams with their bits interleaved—for example, every tenth bit in a high-speed bit stream may be a bit from one subchannel. The bit after that is from a second subchannel and so on, until ten subchannels are derived. This process is called *time-division multiplexing* [1]. The topic is covered in more detail in Chapters 6 and17. We shall frequently refer to the subdivision process, saying, for example, that one pair of wires carries 12 voice channels or that 100 music channels are sent over one television channel.

When frequency-division multiplexing is used, some space is needed between the channels to avoid interference; this is called the *guard band*. For telephone traffic, for example, the world's transmission media are divided into slices of 4 kHz; however, the usable bandwidth is closer to 3400 Hz. If ten 20-kHz hi-fi channels are packed into a bandwidth of 200 kHz, only about 18 kHz of each channel will remain (more than enough for most human ears). The remaining of 2 kHz between each channel is required to separate the channels without causing appreciable distortion.

Time-division multiplexing also is less than 100% efficient, and requires some housekeeping bits among the information bits. Some sawdust is inevitable when a tree is cut into fireplace logs.

DATA   The fastest-growing use of telecommunications
TRANSMISSION   is data transmission. As real-time computer systems improve in design and drop in cost, this use will probably increase even more rapidly.

Data transmission refers to the process whereby computers send data to other computers, and also to transmission between lesser machines, such as telegraph machines, punched card machines, magnetic tape, and disk storages. The most rapidly growing form of data transmission is that in which human beings communicate with a distant computer. They may use typewriterlike devices, terminals that look like a television set and have a keyboard, or Touchtone® telephones. The latter provide the cheapest form of computer input, and the computer responds by generating human-voice words. Such techniques have all kinds of applications, some of which are described in *Telecommunications and the Computer* and *The Computerized Society* [1, 2].

All the items below "telegraphy" in Table 3.1 refer to some form of data transmission. Today telephone and telegraph lines are the transmission facilities most commonly used for this purpose. However, a demand is growing, and will increase, for much faster data transmission channels. It will be increasingly necessary to *dial* a data channel. A person at one location will often want to dial a variety of different computing services. General-purpose terminals in offices will be used by different personnel for different reasons, each dialing the computer he needs.

They are many different ways to construct dial-up data networks. The public telephone network could be used, and it is being used today. Nevertheless, better and cheaper data networks are needed. Some protagonists advocate the building of new networks for data, largely separated from today's facilities. We will explore these arguments in Chapters 8, 9 and 10. Whatever the method employed, it is clear that vastly superior data transmission networks will come into existence in the next decade.

EMERGENCIES   Modern telecommunications offer superb potential for automatically reporting emergencies. Fire warnings or vital machine breakdowns can be signaled instantly to monitoring computers that sound alarms and display required details at fire and breakdown stations. The crime rate is rising at such an alarming rate in the United States (and in other countries with an extensive use of modern communications media) that strong action is becoming increasingly vital [2]. By the end of the 1970s we may find burglar alarms in most homes, coupled to the telephone (or other) cables. The triggering of such alarms would be detected by computers, which again would display relevant information in police stations and would perhaps automatically contact squad cars. If crime in the streets of cities like New York continues its exponential growth, one can imagine pedestrians carrying radio alarms that will enable a computer to dispatch the nearest prowl car to a victim in the street.

The police are already using computers to combat crime [1]. A policeman on his beat or in his car can radio a computer center and obtain immediate information about a suspect he has stopped or a car he is following. Such real-time networks will be nationwide before long and will have vast data banks of police information.

"Emergency communications" in Table 3.1 refers to the capability to preempt normal communications. A designated authority can override normal network controls with an appropriate signal, forcing his way through occupied parts of the network that would otherwise have given him a "busy" signal. He can preempt all or part of the network for emergency purposes.

FOUR PARTS TO TELECOMMUNICATION NETWORKS   The cost of telecommunications can be divided into four main parts:

1. *Long-distance transmission*, the carrying of signals between toll offices. This today consists mainly of wire-pair cables, coaxial cables, and microwave radio transmission between antennas within line of sight on towers, mountain tops, or high rooftops.
2. *Switching*, the elaborate network of switching offices that enable any telephone to be connected to any other telephone and that interconnect some other types of channels as well.
3. *Local loops*, the cables from the subscriber to his local switching office (central office). Telephone and telegraph loops today consist of wire-pair cables. Every subscriber has his own pair of wires to the central office, and nobody else uses these unless he is on a party line. Coaxial cables are now being laid into people's homes for the distribution of television, and these cables have many potential uses other than television (Chapter 5).
4. *Terminals*, the devices the subscriber uses to originate and receive the signals. The vast majority of terminals are telephone handsets. Today, however, an endless array of other devices are being attached to the telephone lines, mainly for data transmission.

In the Bell System, the cost breakdown between these areas is as follows [3]:

long-distance transmission: 17 %

switching: 45 %

local loops: 15 %

terminals: 23 %

COST CHANGES   New technology is changing these costs. The fastest change is in long-distance transmission. The capacity of such systems is increasing greatly without a proportionate increase in cost. Techniques in the laboratory today portend much greater economies of scale in the years ahead. The investment cost of adding a circuit mile to the Bell System now averages $11, and this figure is expected to drop to $1.50 by 1979 [4]. Other estimates suggest that the long-distance bandwidth of new facilities will drop to one tenth of its present cost in the next ten years;

with laser or satellite technology, the cost may drop to one tenth of that figure in the following decade. On the other hand, the signals transmitted will consume greater capacity. Picturephone now needs 250 times the bandwidth of telephone transmission. In addition, satellites are overriding the conventional economics of very long distance transmission in that the cost is independent of distance rather than proportional to it. In the 1970s much larger satellites will be launched, with much more miniaturized electronics, and they will be launched in much greater numbers (Chapter 15).

The other costs are not dropping as fast. The largest portion of the telephone cost is that of switching, which contributes 45 % to the cost of an average telephone call; the figure is 54 % for long-distance calls. Only 28 % of the cost of an average call is in the *switching equipment*; the remainder is in operator salaries [5]. The key to switching economy lies in computers. If used fully, they can reduce the cost in operator salaries. Computers and electronic circuitry, like computer circuitry, are beginning to replace electromechanical telephone exchanges. The cost of computer circuitry is dropping rapidly, and its maintenance costs will be much lower than those of the vast arrays of electromechanical switches. Unfortunately, electromechanical exchanges presently represent billions of dollars of investment and, consequently, will not be replaced quickly. The Bell System, however, estimates that it will be installing computerized exchanges at the rate of one per day by the mid-1970s [6].

The local loops in the telephone network are unlikely to come down much in cost, although the use of concentrators may drop their cost somewhat. A concentrator enables the signals from a number of subscribers to be sent over one pair of wires. It is possible that with digital transmission and time-division multiplexing a quite different configuration of local loops could be built that would be lower in cost. However, the vast investment sums tied up in today's wiring inhibit too drastic a change.

A change that has great potentialities *is* occurring. The local loops are being made to carry a *much* greater bandwidth than telephone loops. This situation resulted from Picturephone development, which required a 1-MHz (million cycles per second) channel into the home. As will be discussed in Chapter 4, the step was achieved by using present telephone wires with frequently spaced amplifiers. Picturephone uses two such channels, one for each direction of picture transmission, in addition to today's voice wires. A 1-MHz channel into a subscriber's location opens up many new possibilities; for example, it could provide a switched data channel of much higher speed than those now in conventional use.

Other channels of enormous capacity are now being wired into homes to carry cable TV. These are not switchable as is a telephone channel; still, new forms of switching could be used in which many subscribers communicate at once over the same coaxial cable. Millimeter-wave radio (explained in Chapter 13) is also being considered for local distribution of television and other signals. A city could have a millimeter-wave "umbrella" supplementing the TV cables and permitting a variety of local communication. Such systems probably will not replace the telephone wires, but they offer new ways of distributing signals other than telephone calls. They will probably be used for television and data transmission and perhaps for "music library" distribution. The problem of local distribution can assume a variety of new forms.

The cost of the telephone set (23% of the total) is the least likely to drop. Instead, many new forms of terminal, some of which are very expensive at the moment, are coming into use. These forms include Picturephone sets, facsimile machines, typewriterlike data terminals, televisionlike data terminals in which the user may have a keyboard or may point at the screen images with a "light pen," terminals using a keyboard and "voice answerback" response, terminals for checking credit cards or obtaining airline tickets, and so on. The cost of data terminals can be expected to drop to perhaps a half or a quarter of today's level as markets increase and logic-circuitry mass-production techniques are perfected. If a true mass market for data terminals develops, it is conceivable that they could drop to costs not much higher than the cost of today's television sets. The domestic television set and data terminal may possibly become combined.

DOMINANT TRENDS   Behind much of the development in telecommunications that will take place in the 1970s, two fundamental changes are occurring. These changes are so powerful that it will be necessary to rethink almost all aspects of the technology. The engineer who rests upon his long experience of telecommunciations technology is having the rug pulled out from under him.

The first of these dominant trends is a continuing and rapid increase in the bandwidths in use. The channel into one's home has long been a 4-kHz telephone channel. Now we see switched Picturephone channels carrying 250 times this bandwidth into the home. Cable TV companies are laying cables with 50,000 times telephone bandwidth into homes, with one cable serving many homes. In 1940 long-distance cables carried 60 telephone conversations. In 1950 the higher bandwidth of the coaxial cable enabled them to carry 600, and in 1960 they carried 3600. At present 17,000 calls are carried by one cable, and the CLOAX cable

(Chapter 14) carrying more than 80,000 will soon be in use. Also ready for trial use is the waveguide, which can carry more than 200,000 voice calls. Beyond that are the laser and optical waveguide systems with a potential bandwidth so high that we do not know how to use it. Radio bandwidths have increased similarly. We have seen in succession the use of long waves (frequencies up to about 100 kHz), the "broadcast" band (around 1 MHz), shortwaves (around 10 MHz), ultrashortwave (around 100 to 1000 MHz), microwaves (up to slightly above 10,000 MHz), and now we are on the threshold of large-scale use of frequencies above 10,000 MHz—millimeter waves.

The other dominant trend is in the use of computers and computerlike logic. Directly accessible computer storages are increasing in size at a staggering rate (Fig. 16.6). Computer logic is rapidly increasing in speed, decreasing in size, and increasing in reliability. Most important, it is fast dropping in cost.

We have now entered a new era in logic technology—the era of *large-scale integration*. What this means in economic terms is that if more than 10,000 of a logic circuit is needed, it can be manufactured at low cost. The process etches the circuit onto a tiny silicon chip in a way that somewhat resembles lithograph reproduction. The process for producing one logic circuit is expensive to set up because the "masks" for etching have to be carefully drawn and debugged. However, once it is set up, additional quantities can be mass-produced like newspaper printing at little extra cost.

In the story that unfolds in the chapters ahead we shall frequently discuss converting signals into a digital form and manipulating the resulting bit stream, sometimes in a complex manner. This process would have been unthinkable ten years ago, but now it appears that it will be increasingly economically viable because of *large-scale integration* circuits.

Increasing bandwidth and decreasing logic cost are a potent combination. Both changes are happening fast. The technical press calls it a "revolution," but we have barely begun to grasp the implications.

REFERENCES

1. James Martin, *Telecommunications and the Computer* (Englewood Cliffs, N.J.: Prentice-Hall, 1969), Chap. 15.
2. James Martin and Adrian Norman, *The Computerized Society* (Englewood Cliffs, N.J.: Prentice-Hall, 1970), Chap. 10.
3. Eugene V. Rostow, "The President's Task Force on Communication Policy," Staff Paper No. 1. Washington, D.C., 1969.
4. *Ibid.*, p. 6.
5. *Ibid.*, p. 44.
6. *Bell System Tech. J.*, October 1969. A complete issue on ESS No. 2.

The first commercial telephone, developed by Bell, went into service in 1877 when a Boston banker leased two instruments. The user placed his mouth and ear to the opening alternately. England had nine telephones installed by the end of 1879. Courtesy AT&T Photo Service.

4 PICTUREPHONE

One of the major changes emerging on the American telecommunications scene is the Bell System Picturephone® service. The Picturephone set now being marketed is shown in Fig. 4.1. The Picturephone service will be integrated into the voice facilities so that customers will be able to dial and use private-branch exchange facilities and Centrex facilities in the same way as with their telephone.

American Telephone and Telegraph forecast that for 1 % of the domestic telephones in service, there will be a Picturephone set by 1980, and the figure will be 3 % for business telephones [1]. This would mean that for 1.5 % of all telephones there would be a Picturephone set by 1980. Picturephone toll calls take more than one hundred times the transmission capacity of telephone on today's channels. On connections to subscribers' premises, Picturephone needs 250 times the bandwidth of telephone. If AT & T meets its forecast, therefore, something like three-quarters of the United States telecommunication bandwidth may be devoted to the Picturephone service by 1980.

Picturephone has been heralded by AT & T as a major social innovation. Julius P. Molnar, executive vice president of the Bell Telephone Laboratories, wrote the following [2]:

> Rarely does an individual or an organization have an opportunity to create something of broad utility that will enrich the daily lives of everybody. Alexander Graham Bell with his invention of the telephone in 1876, and the various people who subsequently developed it for general use, perceived such an opportunity and exploited it for the great benefit of society. Today there stands before us an opportunity of equal magnitude—Picturephone service.

35

Figure 4.1. The Picturephone subscriber uses these three components: (1) A conventional Touchtone telephone. (2) The Picturephone set with its screen, camera, and loudspeaker. (3) A control unit that permits adjustment of the picture, and speaker volume. The control unit also contains a microphone.

He goes on to say in an article in the *Bell Laboratories Record* [2]:

Most people when first confronted with Picturephone seem to imagine that they will use it mainly to display objects or written matter, or they are very much concerned with how they will appear on the screen of the called party. These reactions are only natural, but they also indicate how difficult it is to predict the way people will respond to something new and different.

Those of us who have had the good fortune to use Picturephone regularly in our daily communications find that although it is useful for displaying objects or written matter, its chief value is the face-to-face mode of communication it makes possible. Once the novelty wears off and one can use Picturephone without being self-conscious, he senses in his conversation an enhanced feeling of proximity and intimacy with the other party. The unconscious response that party makes to a remark by breaking into a smile, or by dropping his jaw, or by not responding at all, adds a definite though indescribable "extra" to the communication process. Regular users of Picturephone over the network between BTL and AT & T's headquarters building have agreed that conversations over Picturephone convey much important information over and above that carried by the voice alone. Clearly, "the next best thing to being there" is going to be a Picturephone call.

Not everyone shares this enthusiam. The London *Economist* in a special issue on telecommunications [3] described the use of the Picturephone set as "a social embarrassment" and said that "talking into it was like talking to a mentally defective foreigner." Several corporations in countries other than the United States have developed a similar video-phone, including Plessey of England, Siemen's of Germany, and Toshiba of Japan. However, none of them foresee a public-switched video-phone service in their countries in the near future. The British Post Office expects that such devices might come into service about 1990—twenty years later than the United States. This is not so much a technology gap as a credibility gap. Is it worth paying so much extra to be able to see one's caller? One needs the affluence of the United States and, in particular, the affluence of the Bell System to think so.

Until the late 1970s it seems unlikely that Picturephone cost will be less than ten times that of telephone calls. Later it may drop because of the widespread use of digital lines, the lower cost of logic circuitry, and perhaps because of competition, especially if "foreign attachments" are permitted on Picturephone lines. Possibly it can be sold even at this price. Modern American advertising seems able to sell almost anything. If it saves me business trips on appallingly overcrowded airways or if it enables me to see my grandchildren, it may be a good buy.

Other detractors point out that the picture is not sharp. It is less sharp than television, and a typed document is hardly legible on the screen. If a computer response is transmitted to the caller, it has to be confined to a relatively small number of characters, as in Fig. 4.2, whereas a computer display terminal operating on a dial-up voice line can display more. Nevertheless, there are foreseeable solutions to these problems, as will be discussed later.

Probably the severest criticism from communications men in industry is that the money could have been better spent. Today's voice channels are overloaded. A grave need for better data communication facilities exists. If the same talent, bandwidth, and capital had been spent on a flexible dial-up data network, its effect on the United States economy would have been much greater than that of the Picturephone. Many other countries will have a data network before they have a video-phone service.

Indeed, it is possible that the major economic benefit of the Picturephone service may come from its provision of dial-up lines between major business locations that can carry a million bits of data per second or more. This could revolutionize the way we use computers.

Figure 4.2. Using a Bell System data set, the Picturephone can be linked to a computer and can display data, as shown. In the system illustrated, a user employs his Touchtone buttons to select from a list of services the information he wants displayed. The output of the computer is translated by the data set so that it can appear on the screen. High-speed data devices will be connected to the Picturephone lines in the future, because they provide a dial-up network capable of transmitting 1.3 million bits per second between subscriber premises.

THE SUBSCRIBER EQUIPMENT

The Bell Picturephone System design is one of the most impressive pieces of modern electronic engineering. The subscriber uses three components (Fig. 4.1). First, there is a conventional *Touchtone telephone,* with 12 keys as in Fig. 11.1. Second, there is the *Picturephone set* with its screen, camera, and loudspeaker; and third, there is a *control unit* which contains a microphone and which permits the user to adjust his picture and speaker volume. There is also a separate control unit containing the power supply and line interfacing electronics. This unit is attached to a wall often far away from the Picturephone set so that it does not clutter the user's room.

The main characteristics of the system are listed in Table 4.1. The display unit stands on a nonslip ring and can be turned through almost 360 degrees on it. It can be adjusted for tilt. The screen displays 30 frames per second, composed of about 250 lines per frame. However, the

Table 4.1 PICTUREPHONE STANDARDS

Bandwidth	1 MHz
Screen size	5½″ x 5″
Frames per second	30
Number of lines per frame	250
Normal viewing distance	36″
Normal area of view	17½″ x 16″ to 28½″ x 26″
Close-up area of view	5½″ x 5″

lines are interlaced, with odd-numbered lines on one image and even-numbered lines on the next so as to give a flicker rate of 60 times per second, which makes the flicker virtually unnoticeable (as in television). In a face-to-face conversation we tend to look into each other's eyes. The Picturephone user will usually look at the eyes on the screen rather than at the camera lens, and thus will appear to be looking away slightly. In order to minimize the annoyance of this, the camera lens is placed just above the screen, which is as close as possible to the eyes of the person on the screen. The displacement will make a user appear to be looking down slightly, which we often do in normal conversation.

The camera lens has a viewing angle of about 53 degrees. It is slightly less than 12 inches above the desk, which gives a natural-looking undistorted image of the face of a person sitting at the desk. If the person is 3 feet away, the normal field of view will be 17½ in. × 16 in., which gives the person freedom to move from side to side. The control unit, however, has a SIZE knob that can expand from field of view up to 28½ in. × 26 in., which would permit more than one person to be in view. This "zooming" is done electronically, not by lens movement.

The knob above the camera lens gives a distance setting. It can be set to 3 ft, 20 ft, or 1 ft. At the 20-ft setting, it can view a room or a blackboard. At the 1-ft setting, it views an object lying on the desk underneath it, thus enabling it to project documents or pictures. (In the early system installed at the Westinghouse Electric Corporation, one of the users used to put a photograph of the Westinghouse president in this position whenever the Picturephone rang!) The viewing area under the set is 5½ in. × 5 in. To view this, a swivel-mounted mirror automatically swings in front of the lens, and the picture scanning is appropriately reversed. The top of the field of view is the edge of the ring the set stands on.

The lens iris adjusts automatically to the lighting conditions. The strength of the video signal acts as an "exposure meter," but the upper and lower quarters of the picture are not included in the measure so as to avoid false readings from ceiling lights or white shirts. The iris is a unique friction-free mechanism designed like all of this equipment to be as reliable as possible and not incur maintenance expenses. When the iris is fully open, an automatic gain circuit takes over and the set can be used in quite dim lights. The picture quality, however, is better in good light, and the depth of focus is greater when the iris of the f/2.8 lens is not fully open.

OPERATING
THE EQUIPMENT
A Picturephone call is initiated by the telephone set in exactly the same way as a telephone call except that the # key is pressed prior to the keying of the telephone number. (The # key is the bottom right-hand one of the Touchtone keyboard.) The telephone number keyed will be the regular telephone number of the person in question.

The telephone of the person called makes a distinctive Picturephone ring. The person called picks up the telephone handset and the pictures appear on the screen. The user can adjust his picture with the SIZE, HEIGHT, and CONTRAST knobs. When doing so, he can if he likes look at his own image on his screen by pressing the MONITOR button. If he does not want to appear on the caller's screen, he can press the DISABLE key. He can mute his microphone if he wants with the ON/OFF keys and can adjust the loudspeaker volume of his set with the VOLUME knob.

All the facilities possible with a normal telephone can be used, such as card dialing, key telephone, secretary extensions with or without a picture set, Centrex and private-branch exchange facilities.

TRANSMISSION
ON LOCAL LINES
A major tenet of Picturephone development has been to take maximum advantage of existing Bell System facilities. For this reason, a conventional telephone is used for establishing the calls. The control of the network and the signals used for its control and for dialing are basically the same as for the conventional telephone. What must be added are wires to carry the picture in parallel with the existing local loops for telephones into the subscribers' premises, new switching facilities operated in parallel with the telephone switching but under the same control, and trunks capable of carrying the Picturephone signals as well as today's telephone speech. All these factors have been worked out in a fashion that minimizes both their cost and the consequent unheaval in existing telephone plant.

A bandwidth of 1 MHz (1 million cycles per second) is needed to carry the Picturphone image. This, surprisingly perhaps to those who learned electrical engineering some years ago, is accomplished on wire-pair lines like telephone lines. It was achieved by placing equalizers on the telephone wire pairs about every 6000 ft to give a gain equal and opposite to the attenuation on the wire pair. The line must be equalized over a frequency band from 1 Hz to 1 MHz. Telephone lines now have load coils (inductances) at intervals of about 6000 ft, and the equalizers will re-place these. In this way, Picturephone can be carried by modifying today's telephone connections rather than by replacing them. This is a very important factor in making the service feasible.

The Picturephone set requires two such wire pairs, one for each direc-tion of picture transmission; in addition, the connection to the telephone will remain unchanged. Thus there are six wires from the subscriber's equipment to his local switching office. The equalizers on the picture wires must be carefully adjusted; otherwise distortion accumulating through successive equalizers will produce ghost images caused by echoes in the picture.

For calls that pass through more than one switching office, similar six-wire connections will be used between the offices provided they are not more than about 6 miles apart. Figure 4.3 shows the configuration of these lines.

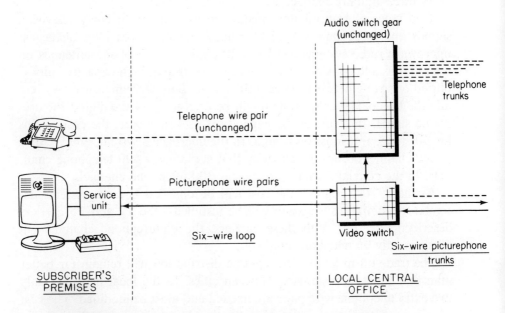

Figure 4.3. Three wire pairs connect the Picturephone equipment to the local central office.

LONG-DISTANCE TRANSMISSION   As will be discussed in Chapter 6, telecommunication signals can be carried in either an analog or a digital form. Digital transmission consists of sending a stream of "bits" like those that flow in computer circuits (Fig. 6.1). Local "loops" from the switching office to the subscriber will mostly remain analog, the signal being carried in its natural form. New technology is making it increasingly economic to transmit signals between switching offices in a digital form, however.

Picturephone signals, as we shall see, can be encoded into a bit stream of 6.3 million bits per second. Eventually all Picturephone transmission between offices (except those less than 6 miles apart) will be done in this digital form. Once digitally encoded, the signals will remain in that form until they reach their destination office or the closest office to it that can decode them (Fig. 4.4).

Transmission in this form has two major advantages: first, it prevents the cumulative distortion that transmission over a variety of analog facilities would cause; second, it is substantially less costly. Picturephone *could* be sent in an analog form over the nation's microwave links; for example, the Bell System TD2 microwave channel could carry one Picturephone signal, just like television. However, the TD2 system can be made to carry a digital signal of 20 million bits per second, and this will carry three digitally encoded Picturephone signals.

Two types of transmission media are used today to carry television signals and large numbers of telephone signals across long distances: microwave radio systems consisting of chains of line-of-site antennas on the hill tops and coaxial cable systems consisting of high-capacity "pipes" buried or strung between poles. Initially the Bell TD2 microwave system and the L4 coaxial cable system will be used to carry the digital Picturephone signal. In terms of the capacity it uses, however, the cost of this long-distance transmission is high. In the microwave system, one Picturephone channel occupies a capacity that could carry 400 telephone channels. In the coaxial system, it displaces 300 telephone channels.

As the 1970s progress, the cost will become less because systems designed for digital transmission will be installed—the T2 and T5 carriers described later on. With these carriers, the Picturephone channel will displace only 96 telephone channels.

The trade-off in local Picturephone distribution will remain far better than that for its transmission between cities. In the local network, only two extra telephone wire pairs are needed and must be modified. The cost of a long-distance call occupying 96 telephone channels will be high; however, certain techniques we will discuss in later chapters promise a dramatic drop in long-distance transmission.

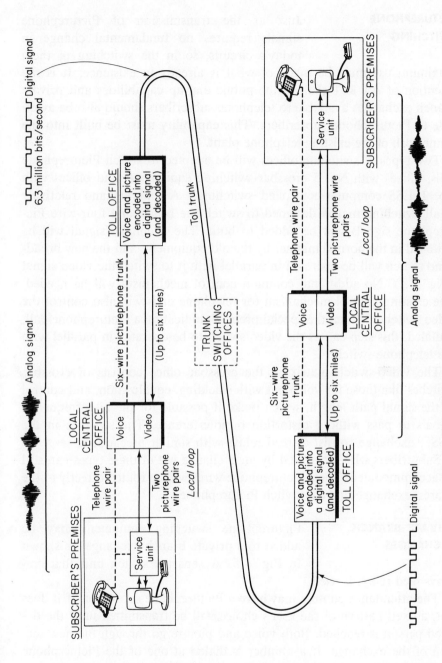

Figure 4.4. The connections used for Picturephone.

PICTUREPHONE SWITCHING   Just as the transmission of Picturephone signals requires no fundamental change in today's circuits, so in the switching of them maximum use must be made of what is already in existence. It is the intention of AT & T that all the public dial-up capabilities and private branch exchanges available to telephone subscribers should also be available to Picturephone subscribers. This capability must be built into the framework of the existing telephone plant.

Two types of switching offices will be adapted to switch Picturephone calls, offices with No. 5 crossbar-switching equipment[2] and offices with No. 1 ESS computer-controlled switching.[3] A separate and relatively small switching matrix designed to switch the broadband four-wire Picturephone signals will be added to both. The telephone signal will be switched in the normal manner by the old equipment, and the new broadband switch will be operated in parallel with it to switch the video signal (Fig. 4.3). No additional common control mechanism will be needed. The existing control mechanism for telephone calls will also control the video switches. When the special prefix #, indicating a Picturephone call, is dialed, this step causes the video switch to be operated in parallel with the telephone switch.

The video-switch matrix in the crossbar office consists of crossbar switches like those for voice but with shielding, equalization, and control of the signal path length, which make it possible for the high-frequency signals to pass without distortion or interference. The switches in the ESS 1 exchange consist of reed relays with similar precautions.

Subscribers who are served by an exchange not of these types can still be accommodated. Their Picturephone wires will be routed directly to the nearest exchange that can switch Picturephone.

PRIVATE BRANCH EXCHANGES   Picturephone switching equipment may be added to a private branch exchange, as shown in Fig. 4.5. A separate switch unit that employs reed relays is used.

The attendant's console may have a Picturephone set or not. If it does not, a fixed pattern of the user's choice will be transmitted until the desired person is reached. Both voice and picture go through the new section of the exchange. If a number is dialed at one of the Picturephone stations and it is preceded by a #, then the line circuit switch will connect it to the Picturephone switchgear and give a busy indication to the tele-

[2]See *Telecommunications and the Computer*, Fig. 18.4 and Chap. 18.
[3]*Ibid.*, Fig. 19.2 and Chap. 19.

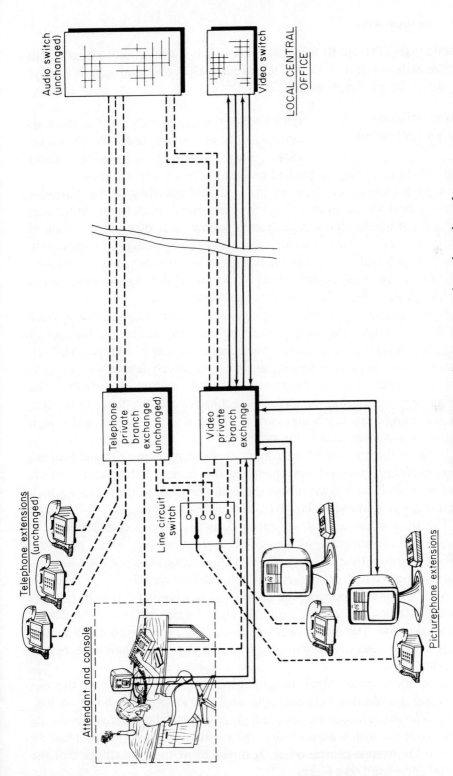

Figure 4.5. The addition of Picturephone equipment to a private branch exchange. *Redrawn from "Video Service for Business," by J. R. Harris and R. D. Williams, Bell Laboratories Record, May/June 1969.*

phone PBX (Private Branch Exchange). If it is not preceded by a #, this switch will connect it to the telephone switchgear and give a "busy" signal to the Picturephone PBX.

IMPROVEMENTS TO PICTUREPHONE   Bell Laboratories are already hard at work on improvements to Picturephone. A set with a color picture is being developed. Better methods of encoding the present picture are being worked upon.

A great increase in efficiency in the digital encoding of the picture is likely to be achieved by encoding the *differences* in each frame rather than the total frame contents. The frames follow each other at intervals of 1/30 of a second on the screen. Normally there is very little change in this brief interval, much less change than in television, where cowboys might be chasing each other across the screen. If the changes alone could be transmitted, the saving would be substantial.

Again, groups of Picturephone signals, digitally encoded, could share the same channel. The picture could be encoded so that the number of bits per second needed to carry it varied with the rate of change. When an almost motionless face is looking at the set, relatively few bits are needed, but if the user turns his head, a large burst of extra bits results. The mechanism for sharing the channels would allocate bit capacity to each set according to its needs and would assume that not all sets had a high degree of picture motion at the same time.

In either there would be occasional overloading of the channel because of momentary excessive movement of the camera subjects. The overloading would result in the movement being blurred. In the uses for which Picturephone is intended, however, this may not matter.

The snag in these forms of encoding is that highly complex logic is needed for encoding the pictures. This logic, however, would be at the toll office rather than at every set, and the cost of complex logic circuits can be expected to drop a great deal in the decade ahead.

Another way of reducing the number of bits needed is to transmit less than 30 frames per second. Probably eight frames per second would be quite adequate. The flicker rate for most of the picture would still be 60. Only in the areas of fast movement would the slow frame rate be detectable.

These ideas could reduce the cost of trunk transmission but not that of the local distribution network (the analog transmission shown in Fig. 4.4). The latter could be reduced greatly, nevertheless, by the use of concentrators, which would lessen the number of wires needed from an area to the nearest central office. It might be at these machines that the digital encoding takes place.

**SLOW
SCANNING**
If Picturephone is going to be a substitute for some business traveling and is going to enable certain people to work at home, a sharper or bigger picture will sometimes be advantageous. Bell Laboratories are working on a slow-scan mode using the same transmission facilities. This will permit high-resolution images to be sent, but it will not be possible to follow rapid movement. The terminal used would have a large screen. Possibly a means for printing documents in this mode will be provided.

**VIDEO
CONFERENCES**
Picturephone in the home is likely to remain too expensive for many would-be users. Public Picturephone booths in the main cities may become popular for talking to far-away friends and relatives.

Higher-quality video equipment will be still more expensive and thus is unlikely to be connected to switched public lines for some time. Instead, it will be used on leased lines, and again city facilities may be available. Some countries that do not plan switched video-phone services in the near future are setting up video-conference rooms in major cities, interlinkable by lines that carry full-quality television pictures. The British Post Office has developed a service called Confravision, in which it provides video-linked studios for businessmen to hold meetings in. It is intended that a Confravision meeting room should exist in most major cities, all interconnectable. Video conferences in industry over private leased lines are also likely to prove valuable. One such link is already in use in the Bell Laboratories between Murray Hill and Holmdel, New Jersey (35 miles). The link is a particularly interesting one, for it uses automatic switching between the conference participants and has features for the transmission of printed material and drawings [4].

Figure 4.6 shows the layout of the conference rooms in this system. Up to nine people sit at a curved table at each conference room location. Three microphones and cameras are used, with up to three persons within range of each. A fourth overview camera has the entire table of participants in view. A fifth camera, which is mounted on the ceiling and points at the center of the tabletop, is used to transmit documents, handwriting, small objects, or drawings, possibly with a speaker's finger pointing out features on them. Cameras can be placed in a variety of other positions, and often a camera giving a close-up of a speaker's face is used.

During normal operation, automatic switching will occur between the cameras. The logic of camera switching has been designed to imitate as closely as possible what a person does in a face-to-face conference. When a person speaks, the camera trained on his face is automatically switched on. The participants at the distant location will thus see his face on the

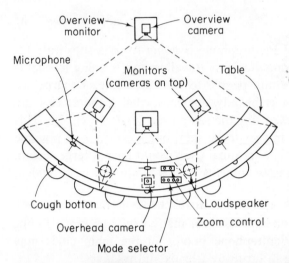

Figure 4.6. The Bell Laboratories video conference system transmits between distant conference locations using ordinary television transmission facilities. In "normal mode" operation the cameras automatically switch to the person speaking.

screens in front of them. When another person speaks, voice-activated switching will switch the transmission to his face. If nobody speaks, the camera will remain on the last speaker for several seconds and the transmission will then be switched to the overview camera.

The screen in front of each participant normally displays the image being transmitted from the distant location. If someone is speaking there, the image will be that of the speaker's face. If no one is speaking, it will be the overview picture. When a person is speaking himself, he will thus see the entire group at the far location. When somebody responds there, he will see a close-up of that person. At the same time, the higher screen

in Fig. 4.6 is displaying the image that is being sent to the far location. This system can be switched by the conference leader into one of three modes other than the normal mode just described.

1. *Locked graphic mode.* Here the outgoing transmission and the screens at the sending location are locked to the overhead graphics camera. A remote zoom control can make this camera show images of objects from 4½ in. x 6 in. to 18 in. x 24 in. The speaker can control the zoom to obtain the clearest image.
2. *Leader graphics mode.* This mode is the same as the locked graphics except that the voice-actuated switching is not suppressed. Sustained speech of a remote participant will cause his face to appear on the screen.
3. *Leader mode.* This mode is used when the conference is being addressed by the leader and the screens contain his face, except when he is interrupted for a sustained period by a speaker from the remote location.

The reaction of the participants to this teleconference scheme was generally very good. One problem had to be dealt with; when persons coughed or sneezed, the loudness of this noise caused switching to the camera trained on them. One can imagine a conference during the annual flu epedemic of the locality in which the screens were filled most of the time with the contorted faces of the various coughers and sneezers. The problem was solved by placing a "cough button" within reach of each participant. Pressing this button prevented the system switching to them when they coughed, sneezed, hic-cupped, or belched.

Teleconference studios may become a common feature of the future. They would enable countries who cannot afford a switched Picturephone service to offer a glamorous facility employing their existing television transmission facilities between cities.

REFERENCES

1. Figures taken from "The President's Task Force on Telecommunications Technology," Staff Paper No. 1, Appendix A, p. 16, Washington, D.C., June 1969.
2. *Bell Laboratories Record*, May–June 1969, introduction to a special issue on Picturephone. Bell Telephone Laboratories, Inc., Murray Hill, N.J.
3. *The Economist*, London, August 9, 1969, a special section on telecommunications.
4. D. Mitchell "Better Video Conferences," *Bell Laboratories Record*, January 1970.

The Bell telephone laboratories gave what was claimed to be the first demonstration of television in 1927. The screen consisted of 50 neon-filled tubes, each divided into small segments. The New York Times stated that television's commercial value was "in doubt."

5 CABLE TV AND ITS OFFSPRINGS

Telephone and Picturephone services provide an individual channel into the home from the switching office. The subscriber has it to himself; unless it is a party line, nobody else uses it. The expense of such a local distribution network is high because of the large number of separate channels needed.

There is another way to transmit information—and that is to link everybody, or large numbers of people, to the same channel. This is done in broadcasting; it is also done in today's cable TV wiring. With modern electronics, however, cable TV could be used in ways quite different from those originally intended, and the potentialities are exciting.

Cable television was started in the 1950s by a local radio dealer in an Appalachian village, a man who was oblivious to its potentialities. He realized that the surrounding mountains were spoiling the reception of television signals, which meant that he would be unable to sell many sets, so he set to work on an enterprising scheme. He erected a tall antenna on a suitable mountain where the reception was good and ran coaxial cable from it into the homes of people willing to pay a small fee. In this way, television came to an Appalachian village, and a technique came into use that should, if allowed, revolutionize home electronics.

Cable television has since been widely installed in rural and metropolitan areas to improve television reception. Figure 5.1 shows the cable passing (unused) through my Manhattan apartment as it now does through many other such apartments in that city. Cable TV is sometimes referred to by the initials CATV (Community Antenna Television).

51

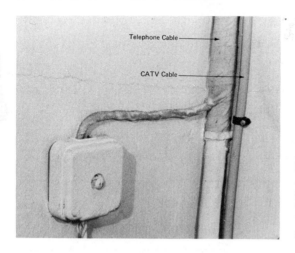

Telephone Cable

CATV Cable

Figure 5.1. The CATV cable in a Manhattan apartment. Many of the television channels in such cables are today unused, leaving a very high bandwidth that could carry other services into the home.

However, with many systems today there is no community antenna; the cable goes directly into the office from which the programs are distributed.

The main sales arguments today are the provision of advertisement-free programs, news, and stock market reports, and the improvement of reception. As the pamphlets frequently mailed to my apartment say:

> Ghosts, unsteady pictures, weird psychedelic colors and assorted eye-jarring interference have plagued TV reception in this area since commercial television first became a reality. . . .
>
> The tall buildings between your set and the broadcasting tower deflect the local TV signals. The signals bounce off these buildings before reaching your antenna. The result: ghosts. And that's just the beginning.
>
> TV signals ricocheting off planes flying overhead are subjected to electrical interference ranging from automobile ignitions to your neighbor's vacuum cleaner. They finally reach your set much the worse for these interferences. You get way-out colors, flip-flopping pictures, snow, herringbone patterns, and assorted other "stray" signals in addition to the signal sent out by the station.
>
> . . . As a cable TV subscriber, you, too, can discover the pleasure of clear, sharp, ghost-free, snowless jitterless television in brilliant color or crisp bright black and white [1].

Cable television has more than five million subscribers in the United States at the time of writing, and the number is increasing rapidly. It has been estimated that by the late 1970s half the homes in this country will be wired for it [2]. There is, however, a wide divergence of opinion as to the rate at which it will spread, as well as its eventual penetration. One body of opinion holds that *all* broadcasting should reach the home by cable, thereby freeing the airwaves for more important uses. It is doubtful whether this situation will occur, although expansion of today's televi-

sion transmission will probably have to be by cable rather than by the consumption of still more of the electromagnetic spectrum. Cable TV detractors say that its usage will be much less. Nevertheless, whether it happens by the mid-1970s or the mid-1980s, it seems probable that half of America's homes will be cable wired in the not-too-distant future. Some enthusiasts say "most" of America's homes.

The little cable in Fig. 5.1 can carry 20 television channels, some not used as yet. Other installations offer 36 and 40 channels, and one organization is trying to obtain a franchise for a 68-channel system. The President's Task Force on Communication Policy estimates that an 81-channel system installed today would cost about $215 per subscriber location. An installation charge of $15–25 per household and a $6 monthly fee might be charged for such a system [3]. With today's technology, this would need two cables. One cable could carry 50 channels.

The American public's appetite for television is enormous. *Business Week* quotes one satisfied cable subscriber as saying "I can see Perry Mason at 180 lb, Perry Mason at 205 lb, and Perry Mason at 230 lb. It just depends on when he made the shows they're rerunning" [2]. However, the public is hardly likely to need 81 broadcast channels. It is in the alternative uses of the cable that the most interesting potentialities lie.

Many of the cable systems in the United States are using some of their channels for services such as weather reports, news headlines, time signals, and stock market reports. Most of these services are provided very inexpensively with a single fixed camera. The cable television companies are starting to run their own films and produce their own programs, in some cases serving a local rather than a statewide need. Local ball games, local shopping information, politicians whose constituency is numbered in a few thousands . . . these seem a natural for cable TV.

A television program need not be expensive. One CATV company set up a very simple studio with little equipment other than a camera and lights [4]. The studio could be used at almost no cost by any group who wanted it. The local school board used this facility to discuss an upcoming bond issue referendum. The program was inexpensive and amateurish, but 40 % of the cable users watched it rather than the major network shows. Only a simple camera is needed for many local events, and a person can learn to operate such a camera very quickly. City council and board meetings, school events, local sports, amateur entertainment productions all seem naturally suited to the local cable. One system in Manhattan transmits Columbia University's basketball games. Thus television becomes a local medium rather than McLuhan's "global" one.

When advertisers discover that local programs have a valuable local audience, advertising patterns may change dramatically. The role of the major networks may also change.

**DIVIDING UP
THE CHANNEL**
The television channel, as we discussed in Chapter 3, has a potentially high information-carrying capability. It gives a higher picture quality than Picturephone because of its higher bandwidth. Cable will probably be the medium that opens up the way for "high-fidelity" television with 5-ft wall screens.

The television channel, like any channel, can be subdivided into channels for signals of smaller bandwidth, which could be used for voice, high-fidelity music, still pictures on the screen, slow-scan images, data, or facsimile. A variety of proposals for future domestic products have arisen from these possibilities.

Many legal problems are connected with the new uses for CATV. The major television networks and the major common carriers are both attacking the cable companies. The FCC, however, has made some moves that seem to favor CATV. In this chapter we shall ignore the turbulent legal questions (discussed in Section V) and discuss the technical feasibility of new uses for the cables.

In practice, a guard band is needed to separate the channels that on cable would be between 10 and 20 % of the signal bandwidth. The telephone companies' "supergroup" for voice channels, for example, has a bandwidth of 240 kHz and carries 60 voice channels, each having a bandwidth of about 3400 Hz. Fifteen percent of the total usable bandwidth is thus lost in the packing process. In packing high-fidelity music channels into a TV channel, let us assume that 20 % is lost. We could then divide the 4.6-MHz television channel into 100 stereophonic music channels, each with a bandwidth of about 37 kHz—enough for very good quality stereophonic reproduction.

**SOUND
CHANNELS**
Of the many television channels, perhaps one could be divided up for classical and one for pop music. The classical music channel could broadcast 100 symphonies, operas, and concertos simultaneously, each occupying a different frequency slot within one television channel. The user would "tune" to the frequency slot he wants. Perhaps a third channel could be divided into nonstereophonic sound channels with a bandwidth of 6 kHz for speech programs. One television channel could carry 600 such sound channels simultaneously (allowing for appropriate

channel spacing). They might be used for continuous news broadcasts, weather forecasts, a time channel (why use a whole TV channel for telling the time?), continuous sports reports, stock market reports, community information reports such as theaters and movies, and shopping news. More than 500 sound channels would still be left for radio plays, talks, poetry, language teaching, and other programs, which might be repeated continuously for each day. Seventy-eight channels on an 81-channel system would remain for Perry Mason and other television! Such is the power of coaxial cable wiring.

As with all we have to say about cable TV, it is far from certain that the full capacity will be utilized. It is far from being utilized today. Moreover, it is uncertain who would pay for such a proliferation of programs, although the programs need not be expensive. There could be frequent reruns of old programs, as well as music without broadcasters. Foreign television could be relayed via satellite. It is worthwhile calculating how the capacity could be used. With Picturephone there is a great shortage of bandwidth; with cable TV there is, at least for today, an embarrassing excess of it. With wall-sized TV screens, however, the excess could soon become used up.

STILL PICTURES   How would one keep track of so many channels? One imagines America's *TV Guide* or Britain's *Radio Times* multiplied a hundredfold! Another way would be to make the network provide its own information. One or more television channels could carry still pictures rather than moving ones. The television picture we watch consists of 30 pictures per second. If the channel carried separate still pictures, these could also be carried at the rate of 30 per second. Keys on the user's set would instruct it which "page" to display, and enough "pages" could be carried to give all of the program information.

For still pictures, the television channel could be divided up timewise or frequency-wise. If divided up by frequency, the channel could be split into 200 subchannels and each of 200 pictures would be scanned every 10 seconds. The receiving set would have some means for storing a picture and regenerating the image on the screen. Again the user would "tune" to the frequency band he required. The "response time" or the time he would have to wait between selecting a picture and completely receiving it would be 10 seconds. If a response time of 50 seconds were tolerable, then 1000 subchannels could be used, giving access to 1000 pictures.

If the channel is subdivided by time rather than frequency, the pictures would be scanned and transmitted at the same rate as conventional television. The difference would be that when the set has received one frame, it keeps it, instead of immediately changing it, as today's television set does. Again, the set would need some means of storing the frame. Now, instead of a mechanism for tuning to a given frequency band, the set must have an accurate timing mechanism for picking the correct frame. One or more of the frames sent in each sequence would be a frame solely for synchronization purposes.

The still-picture facility would have many uses, and in a system with 50 or so channels it would seem likely that several of the channels could be still picture channels. Again, news services, weather forecasts, and stock market reports could use some of the frames in the channel. One's daily newspaper in the future may occupy one such television channel. Local shops, theaters, and movies may use such a channel in color, and many other local advertisers could employ it. Programmed instruction sessions may be scheduled, with the student "branching" from one page to another according to his level of knowledge or success in answering questions.

It has been proposed that facsimile machines should become part of such equipment in the future, and the user would have the facility to copy pages he wanted.

TYPES OF RECEIVING SETS   A cable carrying the foregoing types of signals could have a wide variety of different types of sets attached to it, not all of which would be expensive. It could have a simple television set, black and white or color, as today. It could have a sound receiver and no television, either an inexpensive set or the best of today's high-fidelity equipment. A somewhat more expensive machine would have the still-picture facility with perhaps a remote keyboard held on one's lap.

So far we have discussed only analog signals like sound and television. If we allow some channels of the cable to carry high-speed digital signals like a computer, then many more possibilities arise. Another, somewhat more expensive category of sets would handle digital operations like a computer terminal.

AN ANSWERBACK CHANNEL   The transmission in the preceding examples was entirely one way, from a broadcasting station to the home, as with present television and radio. A cable rather than an airwave system, however, offers the possibility of the subscriber being able to respond. His response would be interpreted by a computer. If he can make responses, many new uses of the medium become possible. Let us suppose that one cable serves a maximum of about 30,000 sets. If the number of sets in a community exceeds this amount, more than one cable will be needed. A second cable would not increase the distribution cost per set by a significant amount. One of the television channels could carry a continuous digital bit stream. The television bandwidth could carry a stream of 6 million bits per second. Let us suppose that the cable is arranged in a loop so that this bit stream flows continuously. At the transmission station, the bit stream enters a computer. Its speed is less than that of modern high-speed computer channels. The bit stream is used for carrying information from the subscribers' sets to the computer.

The bit stream could, for example, inform the computer whether a set is switched on and which channel or subchannel it is tuned to. Billing according to usage could be organized in this way. Detailed statistics about the system usage and about advertisement coverage could be collected. More interesting, the computer could read the setting of manually operated keys, and hence the system could be used to solicit subscriber responses.

Clearly there are many ways in which the bit stream could be organized for this purpose. Let us examine a detailed proposal in order to assess the potentiality:

The bit stream is composed of six-bit characters. It thus transmits one million characters per second. The sets are arranged for addressing purposes in groups of eight. The sets in the group need not be in one location or in any way related; they could be miles apart. A maximum of slightly less than 4096 (2^{12}) groups of sets can be connected to one cable loop; hence two six-bit characters can be used for addressing the sets.

The characters flowing on the loop are organized into "blocks", and every block relates to one group of eight sets. The composition of a block is shown in Fig. 5.2. It has space for eight characters, one from each set in the group. If a set has a character to send, it waits until a block with the address of its group arrives and inserts the character in the slot for that set.

The first four characters of the block are a synchronization pattern. These are four unique synchronization characters. This character cannot be used for any other purpose. The synchronization pattern is continuously checked to ensure that the sets insert their bits in the correct place in the bit stream.

The fifth and sixth characters are 12 bits giving the address of the group of sets (from 0 to 2^{12}, but the synchronization characters cannot be used, so $2^{12} - 2 \times 2^6 = 3698$ groups of sets are permitted, giving 31, 744 sets in total on one cable.)

The last two characters are an error-checking code. A polynomial code that gives an extremely high measure of protection from damage to the bits caused by noise on

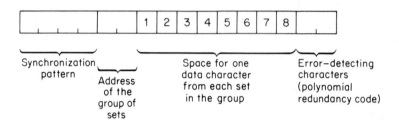

Figure 5.2.   Bit pattern for carrying data from the sets.

the cable would be used. These two characters check both the group address and the data. If an error is detected, then a code will be placed in the character slot for that set, telling the set to re-send its previous character.

The television channel can transmit 62,500 such blocks per second. As there may be 3968 groups of sets, each set receives 62,500/3968 = 15.8 blocks per second. In other words, this scheme can transmits 15.8 characters per second from each set. This is sufficient for a fast typist. IBM's typewriterlike terminals transmit 14.8 characters per second.

The user could have a variety of different types of keyboard for responding. He might have a typewriter keyboard or a keyboard similar to the Touchtone telephone. He might simply have two keys that say Yes and No. It is assumed that the circuit for inserting the characters he sends into the bit stream would be made using *large-scale integration* (discussed in Chapter 16). It would be mass-produced in large quantities and hence be relatively inexpensive.

A TWO-WAY   The digital response channel has far more uses
DIGITAL   if the subscriber can enter into some form of
CHANNEL   dialogue with this new medium. The answer-
back to the subscriber could come either in the form of selected still pictures or in the form of a digital reply, as at a computer terminal. Still-picture responses are attractive for certain applications, but the channel has a limited capacity for them, as we shall discuss shortly. Digital channels in both directions, however, give great flexibility, and any number of the 30,000 subscribers on the cable could use them simultaneously.

Again, there are many ways in which two-way digital transmission

could be organized. Let us again make a specific proposal in order to examine the potentialities quantitatively.

Suppose that a second television channel is used to carry a digital bit stream, this time in the opposite direction. A million six-bit characters flow *to* the sets on this channel from a computer. The characters are again organized into "blocks." This time a block consists of 128 characters; its composition is shown in Fig. 5.3.

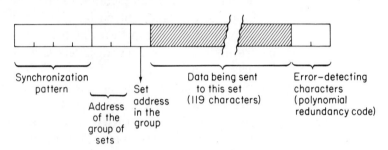

Figure 5.3.   The bit pattern for carrying data *to* the sets.

The block starts with four synchronization characters, followed by the group address. Now, however, the block is addressed to a specific set, and so the seventh character is the set address. The block can carry 119 characters of data to that set. The last two characters are once more used for error detection.

The television channel can transmit 7812.5 such blocks per second. If it transmits to each group of sets in sequence as on the input channel, each group will receive a response every $3968/7812.5 = 0.51$ second. Normally each set will thus receive a response time of up to half a second from the cable. The mean response time will be a quarter of a second. It is improbable that two of the eight sets in a group would be receiving a response in the same half-second. If such is the case, however, one of them will have to wait for the next block and so will have a response time between half a second and one second. The worst possible response time is 4 seconds, but this is very unlikely. If a screen is being filled with a message longer than 119 characters, the data will be sent in successive groups of 119 characters following each other at half-second intervals. Similarly, if a printer, like a computer printer, was used on the wire, it could operate at 120 lines per minute.

We have thus employed two channels of our television cable to enable us to attach terminals like those on a computer system. The domestic television set can become a computerlike terminal. Thirty thousand such terminals can be connected to one TV cable and can operate simultaneously. The input channel keeps up with high-speed typing, and the output channel normally gives a half-second response time.

The computer controlling such a loop would be likely to be connected to other computers carrying on a wide variety of specialized functions. It would pass on the messages from subscribers via the public telephone network or over leased telecommunication lines. In this way, the subscriber would be able to contact any computer systems that are available for public use.

DIALOGUE
WITH PICTURES

As discussed earlier, a cable TV channel can be made to carry still pictures. Many appealing forms of dialogue can be imagined in which the responses come in the form of a color picture on the TV screen rather than in the form of a digital response (with letters and digits).

Consider the following case as an illustration of the possibilities. A woman wants to continue planning her vacation, or perhaps she dreams about a vacation she is unlikely to have. She uses her set to contact the local travel-agent computer, having been instructed how to do so by advertisements on the other channels. The travel-agent computer has her account record, which states how previous dialogues have concluded. It displays the following message on her screen:

> GOOD EVENING MRS. SMITH.
> WOULD YOU LIKE A CONTINUED PRESENTATION
> ON ONE OF YOUR PRIME SELECTIONS:
> 1. PERU?
> 2. BOLIVIA?
> 3. HAWAII?
> 4. EASTER ISLAND?
> IF SO PLEASE KEY THE ABOVE NUMBER

She presses the 4 key on her Touchtone telephonelike keyboard. The "4" travels over the digital input channel, via the cable computer to the travel-agent computer. The travel-agent computer responds:

> EASTER ISLAND.
> WHICH OF THE FOLLOWING WOULD YOU PREFER:
> 1. SCENERY?
> 2. ENTERTAINMENTS?
> 3. HOTEL INFORMATION?
> 4. SHOPPING?

5. TIME-TABLE?
6. TO MAKE RESERVATIONS?
   PLEASE KEY THE ABOVE NUMBER.

She presses the 1 key and the screen says:

SCENERY.
PLEASE PRESS THE # KEY TO CHANGE SLIDES AND BEGIN PRESENTATION.

She presses the # key. There is a pause of about 6 seconds, and then a magnificent aerial view of Easter Island appears on the screen in color. The woman looks at breathtaking pictures for the next half hour and then presses a key labeled END. A message appears on the screen trying to sell her a booking. She responds negatively and switches the set off.

The pictures may come from 35 mm slides in a large random-access file, or from frames on microfiche or EVR cartridges in units controlled by the travel-agent computer. The image is scanned and passed over a communication line to a switch unit at the head of the TV cable and from there was switched to the appropriate subchannel.

All manner of applications for dialogue with pictures can be imagined. It may be particularly valuable for teaching, especially with children. Madison Avenue probably would be the richest source of material for such a channel initially, but other longer-lasting "programs" would steadily amass like books in a library.

Again, there are a variety of ways in which such a channel could be organized. Perhaps the simplest way to organize it would be to assign a still-picture subchannel permanently to a subscriber for the duration of his picture dialogue.

Let us again assume that one cable is limited to 30,000 subscribers. At the peak-viewing period, not more than 40 % of the subscribers are watching their sets. Most subscribers still spend most of their time with "passive" channels rather than with those demanding a response. In many homes the passive channels are a half-perceived background to daily living. During the peak-viewing period, less than 2 % of the viewers, say, will use the still-picture channel—say 200 viewers per cable.

One television channel could be used for 200 viewers and give a response time of 10 seconds with frequency-division multiplexing and from 0 to 10 seconds with time-division multiplexing. If two channels were used for the 200 viewers, these times would be halved.

Such a system would have a small but finite probability that all 200 subchannels are busy when a subscriber attempts to obtain a still-picture service. In this case, he will obtain a "busy" indication, just as he sometimes does from a telephone exchange or a tie-line system. If the average viewer watches the still-picture channel for no more than half an hour at a time, another channel will become free, on the average, every 7.5 seconds. If the subscriber attempts to obtain a channel again a minute later, he will probably succeed unless a higher than normal demand is building up. A message may be flashed on the screen to this effect.

If the demand for still-picture service is in fact greater than the preceding figures indicate, more than one of the many television channels may be used for it. Another way to solve this problem would be to send a set address with each still picture. The set only displays the picture addressed to it and that picture does not then continue to cycle round the cable. This practice would give a much more efficient use of the still-picture channel but may result in a higher set cost.

The response times in practice might be much lower than the above figures, and the probability of delay very low, because far fewer than 30,000 sets would be attached to each cable.

MUSIC LIBRARY

In the preceding scheme, the user, in conjunction with various remote computers, is *selecting* the picture he wants from a potentially large number of pictures. The same user-selection process could be applied to music.

Let us suppose that a music "library" that plays an extremely large number of pieces of music continuously is set up. A tape deck, like a videotape machine, might be used to play 100 pieces of stereophonic high-fidelity music continuously. Forty such machines might be used with different playing times to cater to the different lengths of the works, which would mean that 4000 music tracks were playing simultaneously. The music would include every major symphony, opera and concerto, and a vast amount of jazz, pop, and show music. The cost of the multichannel tape deck might be $4000, giving a cost for the entire facility of perhaps $250,000. One coaxial cable could carry all 4000 sound channels to locations where they would be appropriately switched on to the TV cable subchannels. If one such facility served a million subscribers, the cost per subscriber would be very low.

The majority of sound channels on the cable would carry music determined by the programming authorities. The majority of subscribers, most of the time, would be content to listen to what was fed to them, especially if a large number of channels existed, as indicated earlier. At any one time, a minority of subscribers would be playing a symphony or other piece of music *requested* from the library. Let us again suppose that of the 3,000 subscribers, not more than half have their sets switched on in the peak hour. Of these subscribers, not more than 20% are likely to be listening to sound-only programs, and of those that are, at least 90% are doing so in a passive way rather than on a *request* basis. The result is a maximum of 300 people using request channels. The number of stereo-phonic hi-fi channels that can occupy one television channel is 100, and so three such channels would be needed. Three television channels would probably be enough for the music request service, particularly if a small charge is made, say 25¢, for every hour of request listening.

Let us suppose that on an average day 1% of the subscribers use the music library service at some time, for an average of 2 hours per day. The 25¢ charge would bring in $50,000 per day ($18.25 million per year) from the one million subscribers connected to one music library. This figure would give a large profit margin. Such a system would seem to be economically feasible in large cities. In smaller cities, use would depend on the cost of long-distance transmission of the television channel from a remote music library.

In addition, it seems possible that in the United States some advertising might be added at the end of a piece of music, to fill up the gap between the end of the music and the end of the tape. The advertising might be designed to appeal to music lovers or hi-fi enthusiasts. A highly selective market for certain products may exist. Market research, for example, might discover that persons who listen to Götterdämmerung are more likely to take aspirin than the average consumer and aspirin could be advertised with that tape!

SELECTION OF OTHER PROGRAMMING The idea of being able to request one's programs instead of merely accepting broadcasts, which are usually planned for a mass audience rather than a selective audience, is appealing. With coaxial cables going into a large proportion of homes, this step seems economically feasible with sound-only programs. It does not appear feasible with TV programs, given today's equipment costs. A TV program needs 120 times the bandwidth of high-fidelity stereophonic music—600 times the bandwidth of lesser-fidelity monaural sound. If 25¢ an hour is an economically viable charge for sound, it would be necessary to charge many times this amount, perhaps 40 times or more for television. Also,

Figure 5.4   The EVR (Electronic Video Recording) System of CBS.

In EVR, color television signals are encoded on a reel of black-and-white film images, packaged in a disk seven inches in diameter (below). The user drops the disk onto a spindle on a player (right) which is connected to a domestic television set. The disk provides up to 25 minutes of playing time (30 minutes with European standards). If black-and-white television is used the disk provides twice this running time. Color films consist of two sets of frames (opposite page), "luminance" images which look like normal black-and-white film, and "chrominance" images which contain the information necessary to add the color.

Photographs Courtesy of Columbia Broadcasting System

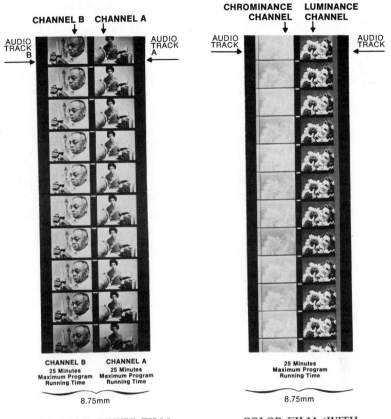

BLACK-AND-WHITE FILM
(TWO CHANNELS).

COLOR FILM (WITH
STEREOPHONIC SOUND).

The film runs at 60 frames per second (50 in Europe). It can be stopped at any point to show a still frame. The disk contains 180,000 black-and-white frames or 90,000 color ones. This is the equivalent of about 500 library books. If the mechanism were used in a computer-assisted instruction terminal, the remote computer could select the appropriate frame by means of encoded numbering on the magnetic (sound) track. The potential of this technology is immense.

whereas 300 request sound channels could be accommodated on today's TV cables, 300 television channels could not.

Another possibility that has been suggested is to send movies to the home on request at off-peak viewing hours and store automatically on videotape. Suppose that 20 television channels are used for this purpose during 16 hours of the day (thus avoiding an 8-hour peak-viewing period). If the average length of a movie is 2 hours, then 160 movies per day could be transmitted to subscribers whose tape machines would be automatically switched on when the movie began.

The economics of this project, however, seem less attractive than the music library scheme. The subscriber would need an expensive automatically controlled video-recording machine. Suppose that 1 % of the subscribers are prepared to request one movie per week at a cost of $10. The total would be $5.2 million per year for one million subscribers.

The mass-marketing by CBS, RCA, and other organizations of cartridge-recorded movies that can be played over a conventional television set makes the prospects of selected movie transmission look unpromising.

Home selection of movie transmission does not appear likely in the next ten years. After that, who knows which way the market will go? Cartridges will remain expensive to buy and rental services may not have consumer appeal—record libraries have had little use. Further, *interactive* movies such as computer-assisted instruction with movie sequences or other man-machine dialogue schemes with movies, may become the basis of a consumer market using the TV cable.

INTERCONNECTIONS TO THE CABLE
At the cable head, the various channels will be connected to other forms of transmission media. The television signals may come from local cables to television studios or from long-distance links such as those provided by the Bell System for today's television distribution. They may come from satellite links, often with other countries. In the future they will sometimes come from direct-broadcast satellite antennas (see Chapter 15). The sound channels may go through switching facilities to sound studios or a music library. They may also be switched to long-distance links or obtained from radio antennas.

The digital channels, when used, will enter and be transmitted from a computer at the cable head, which, in many cases, would act as a means of switching the data to links with other distant, special-purpose computers. The computer would permit a terminal on the cable to "dial" computers anywhere and make the connections over the public network.

Millimeter-wave radio systems that can connect to TV cables have been developed. The present systems operate in the 18-GHz frequency range and transmit 12 television channels simultaneously over distances up to 6 miles. A 100-MHz bandwidth is used. This system is useful for by-passing natural cable barriers like rivers and for reaching outlying subur-ban areas. There are some difficulties with millimeter wave radio as we shall discuss in Chapter 13. Some success has also been had with infrared transmission.

REFERENCES

1. Advertisement from Manhattan Cable Television.
2. "Cable TV Leaps into the Big Time," *Business Week*, November 22, 1969.
3. The President's Task Force on Communication Policy, Staff Paper No. 1 Eugene V. Rostow, "A Survey of Telecommunications Technology," Washington D.C., June 1969.
4. *Ibid.*, Appendix G, p. 55.

1845. Samuel F.B. Morse sending his first public telegraph message: "What hath God wrought."

6 ANALOG VERSUS DIGITAL TRANSMISSION

Basically, there are two ways in which information of any type can be transmitted over telecommunication media: analog or digital.

Analog means that a continuous range of frequencies is transmitted. The sound you hear and the light you see consist of such a continuous range. Sound, as any hi-fi enthusiast knows, consists of a spread of frequencies from about 30 to 15,000 Hz or, for people with very good ears, 20,000 Hz. It cannot be heard below 30, and it cannot be heard above 20,000. If we wanted to transmit high-fidelity music along the telephone wires into your home (which is technically possible), we would send a continuous range of frequencies from 30 to 20,000. The current on the wire would vary *continuously* in the same way as the sound you hear.

The telephone company, being very concerned about money, transmits a range of frequencies that may vary from about 300 to 3400 Hz only. This is enough to make a person's voice recognizable and intelligible. When telephone signals travel over lengthy channels, they are packed together, or *multiplexed*, so that one channel can carry as many such signals as possible. To do this your voice might have been raised in frequency from 300–3400, to 60,300–63,400 Hz. Your neighbour's voice might have been raised 64,300 to 67,400. In this way, they can travel together without interfering with one another; but both are still transmitted in an *analog* form—that is, as a continuous signal in a continuous range of frequencies.

Digital transmission means that a stream of on/off pulses are sent, like the way in which data travel in computer circuits. The pulses are referred to as *bits*. It is possible today to transmit at an extremely high bit rate.

Figure 6.1 shows an analog signal and a digital signal. A transmission

An analog signal

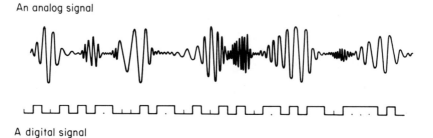

A digital signal

Figure 6.1.   Any information can be transmitted in either an analog or a digital form.

path can be designed to carry either one or the other. As we shall see in the chapters ahead, this fact applies to all types of transmission paths— wire pairs, high-capacity coaxial cables, microwave radio links, satellites, and the new transmission media, such as waveguides and lasers.[1]

In order to follow the arguments in the chapters ahead, it is important to understand that *any type of information can be transmitted in either an analog or a digital form.*

The telephone channel reaching our home today is an analog channel, capable of transmitting a certain range of frequencies. If we send computer data over it, we have to convert that digital bit stream into an analog signal using a special device known as a *modem*. This converts the data into a continuous range of frequencies—the range as the telephone voice. In this way, we can use any of the world's analog channels for sending digital data.

On the other hand, where digital channels have been constructed, it is possible to transmit the human voice over them by converting it into a digital form. Similarly, *any* analog signal can be digitized for transmission in this manner. We can convert hi-fi music, television pictures, temperature readings, the output of a Xerox machine, or any other analog signal into a bit stream. High-fidelity music would need a larger number of bits per second than telephone sound. Television would need a much higher bit rate than sound transmission. The bit rate needed is dependent on the

---

[1]These different types of transmission media are described in the author's *Telecommunications and the Computer*, Chap. 8.

bandwidth, or range of frequencies, of the analog signal, as well as the number of different amplitude levels we want to be able to reproduce.

Almost all the world's telephone plant today grew up using analog transmission. Much of it will remain so for years to come because of the multi-billions of dollars tied up in such equipment. However, digital technology is rapidly evolving, and major advantages in digital transmission are beginning to emerge. It is probable that if the telecommunication companies were to start afresh in building the world's telecommunication channels, there would almost entirely be digital with the exception of the local "loops" between a subscriber and his nearest switching office. A new and different form of plant would be installed using a technique called *pulse code modulation*, in which the voice and other analog signals would be converted into a stream of bits looking remarkably like computer data. Some developing countries with a less-massive investment in old equipment are installing pulse-modulated systems, and lines of this type are already in extensive use in the United States.

If digital bit streams form the basis of our communication links, then computer data will no longer need to be converted into an analog form for transmission, as it is today. However, analog information, such as the sound of the human voice, needs to be coded in some way so that it can be transmitted in the form of pulses and then decoded at the other end to reconstitute the voice sounds. This is already done on many short-haul telephone trunks. The circuits of the future will be designed to transmit very high speed pulse trains into which voice, television, facsimile, and data will all be coded and sent in a uniform manner. Instead of manipulating data so that they can be squeezed into channels designed for voice, the voice will be coded so that it can be sent over channels that are basically digital.

ECONOMIC
FACTORS

The economic circumstances favoring digital transmission stem from two main factors. First, it is becoming possible to build channels of high bandwidth—that is, high information-carrying capacity. Indeed, it is now appreciated that many existing wire-pair channels, which represent an enormous financial investment, could be made to carry much more traffic than they are currently doing. However, a high level of multiplexing is needed to make use of high-capacity channels. In other words, many different transmissions must be packed together to travel over the same channel. When such packing occurs, the circuit-mile cost for one signal, such as a telephone transmission, drops greatly.

But the cost of packing and unpacking them, plus switching them, remains high.

When analog signals are multiplexed together, each must occupy a different range of frequencies within the overall range that is transmitted. The frequency range (bandwidth) available is divided up and allocated to the separate signals. This process, known as *frequency-division multiplexing*, uses fairly expensive circuit components such as filters. When thousands of telephone conversations travel together over coaxial cable or microwave links, they must be demultiplexed, switched, and then multiplexed together again at each switching point. While there is great economy of scale in the transmission, there is not in this multiplexing and switching operation. As the channel capacities increase, so the multiplexing and switching costs assume a greater and greater proportion of the total network cost.

Digital circuitry, on the other hand, is dropping in cost at a high rate. With the maturing of large-scale integration techniques, it will drop even more. Where digital rather than analog transmission is used, this increasingly low-cost circuitry will handle the multiplexing and switching. The telecommunication networks will become in some aspects like a vast digital computer.

There is one other major advantage in using digital techniques for transmission. In analog transmission, whenever the signal is amplified, the noise and distortion is amplified with it. As the signal passes through its many amplifying stations, so the noise is amplified and thus is cumulative. With digital transmission, however, each repeater station regenerates the pulses. New, clean pulses are reconstructed and sent on to the next repeater, where another cleaning-up process takes place. Therefore the pulse train can travel through a dispersive noisy medium, but instead of becoming more and more distorted until eventually parts are unrecognizable, it is repeatedly reconstructed, and thus remains impervious to most of the corrosion of the medium. Of course, an exceptionally large noise impulse may destroy one or more pulses so that they cannot be reconstructed by the repeater stations.

A distinctive characteristic of digital transmission is that a much greater bandwidth is required. In order to send a given quantity of telephone conversations, for example, we would need a much higher bandwidth than with the analog systems in use today. However, because the signal is regenerated frequently, the pulse code modulation signal can operate with a lower signal-to-noise ratio. Thus there is a trade-off between bandwidth and signal-to-noise ratio in the transmission of a given quantity of information. If a given pair of wires is used, for example, a wider range of frequencies can be employed for transmission because of the frequent regeneration of the signal, and because only two states of a binary signal need to be detected—not a continuous range of amplitudes as in an analog signal.

An additional economic factor is the rapidly increasing use of data transmission. Although data transmission still employs only a small proportion of the total bandwidth in use, it is increasing much more rapidly than other uses of the telecommunication networks. Data are transmitted much more economically over a digital circuit than an analog circuit. With the present state of the art, ten times as much data can be sent over a digital voice line as over an analog voice line.

Thus four factors are swinging the economics in favor of digital transmission.

1. The trend to much higher bandwidth facilities.
2. The decreasing cost of logic circuitry, which is used in coding and decoding the digital signals and in multiplexing and switching them.
3. The increase in capacity that results from the use of digital repeaters at frequent intervals on a line.
4. The rapidly increasing need to transmit digital data on the networks.

In terms of the immediate economics of today's common carriers, pressed for capacity, digital transmission is appealing for short-distance links because with relatively low-cost electronics it can substantially increase the capacity of existing wire pairs. This is particularly important in the congested city streets.

An important long-term advantage is the fact that all signals—voice, television, facsimile, and data—become a stream of similar-looking pulses. Consequently, they will not interfere with one another and will not make differing demands on the engineering of the channels. In an analog signal format, television and data are much more demanding in the fidelity of transmission than speech and create more interference when transmitted with other signals. Eventually, perhaps, there will be an integrated network in which all signals travel together digitally.

We will say more about the detailed economics in the chapters which follow.

PULSE
AMPLITUDE
MODULATION
In order to convert an analog signal such as speech into a pulse train, a circuit must sample it at periodic intervals. The simplest form of sampling produces pulses, the amplitude of which is proportional to the amplitude of the signal at the sampling instant (see Fig. 6.2). This process is called *pulse amplitude modulation* or PAM.

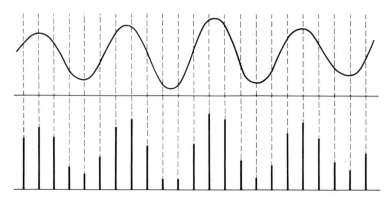

Figure 6.2. Pulse amplitude modulation (PAM).

The pulses produced still carry their information in an analog form; the amplitude of the pulse is continuously variable. If the pulse train were transmitted over a long distance and subjected to distortion, it may not be possible to reconstruct the original pulses. To avoid this we employ a second process, which converts the PAM pulses into unique sets of equal amplitude pulses so that we need only detect the presence or absence of a pulse (bit), not its size. As we shall see in Chapter 17, the PAM pulses themselves are used in certain switching equipment in which the switching is done by electronically controlling the flow of PAM pulses.

PULSE CODE
MODULATION
The amplitude of the PAM pulse can assume an infinite number of possible values ranging from zero to its maximum.

It is normal with pulse modulation to transmit not an infinitely finely divided range of values but a limited set of specific discrete values. The input signal is *quantized*. This process is illustrated schematically in Fig. 6.3. Here the signal amplitude can be represented by any one of the eight values shown. The amplitude of the pulses will therefore be one of these eight values. An inaccuracy is introduced in the reproduction of the signal by doing this, analogous to the error introduced by rounding a value in a computation. Figure 6.3 shows only eight possible values of

① The signal is first "quantized" or made to occupy a discrete set of values

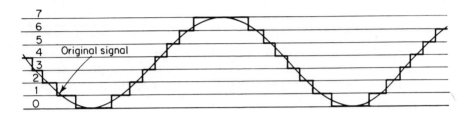

② It is then sampled at specific points. The PAM signal that results can be coded for pulse code transmission

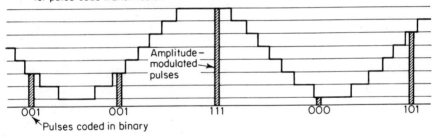

③ The coded pulse may be transmitted in a binary form

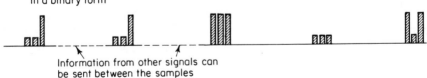

Information from other signals can be sent between the samples

Figure 6.3.   Pulse code modulation (PCM).

the pulse amplitude. If there were more values, the "rounding error" would be less. In systems in actual use today, 128 pulse amplitudes are used, or 127 to be exact for the zero amplitude is not transmitted.

After a signal has been quantized and samples taken at specific points, as in Fig. 6.3, the result can be coded. If the pulses in the figure are coded in binary, as shown, three bits are needed to represent the eight possible amplitudes of each sample. A more accurate sampling with 128 quantized levels would need seven bits to represent each sample. In general, if there were N quantized levels, $\log_2 N$ bits would be needed per sample.

The process producing the binary pulse train is referred to as *pulse code modulation*. The resulting train of pulses passes through frequent repeater stations that reconstruct the pulse train, and is impervious to most types of telecommunications distortion other than major noise impulses or dropouts. The mere presence or absence of a pulse can be recognized easily even when distortion is present, whereas determination of pulse magnitude would be more prone to error.

On the other hand, the original voice signal can never be reproduced exactly, because of the quantizing errors. This deviation from the original signal is sometimes referred to as *quantizing noise*. It is of known magnitude and can be reduced, at the expense of bandwidth, by increasing the number of sampling levels; 128 levels, needing seven bits per sample, are enough to produce telephone channels having a signal-to-noise ratio comparable to that achieved on today's analog channels.

HOW MANY SAMPLES ARE NEEDED? The pulses illustrated in Figs. 6.2 and 6.3 are sampling the input at a limited number of points in time. The question therefore arises: How often do we need to sample the signal in order to be able to reconstruct it satisfactorily from the samples? The less frequently we can sample it, the lower the number of pulses we have to transmit in order to send the information, or, conversely, the more information we can transmit over a given bandwidth.

Any signal can be considered as being a collection of different frequencies, but the bandwidth limitations on it impose an upper limit to these frequencies. When listening to a violin, you hear several frequencies at the same time, the higher ones being referred to as "harmonics." You hear no frequencies higher than 20,000 Hz, however, because that is the upper limit of the human ear. (The ear has a limited bandwidth like any other channel.) When listening to a full orchestra, you are still hearing a collection of sounds of different frequencies, although now the pattern is much more complex. Similarly, other signals that we transmit are composed of a jumble of frequencies. A digital signal can be analyzed by Fourier analysis into its component frequencies.

It can be shown mathematically that *if the signal is limited so that the highest frequency it contains is W hertz, then a pulse train of 2W pulses per second is sufficient to carry it and allow it to be completely reconstructed.*

The human voice, therefore, if limited to frequencies below 4000 Hz, can be carried by a pulse train of 8000 PAM pulses per second. The original voice sounds, below 4000 Hz, can then be *completely* reconstructed.

Similarly, 40,000 samples per second could carry hi-fi music and allow complete reproduction. (If the samples themselves were digitized, as with PCM, the reproduction would not be quite perfect because of the quantizing error.)

Table 6.1 shows the bandwidth needed for four types of signals for

Table 6.1 BANDWIDTHS AND EQUIVALENT PCM BIT RATE FOR TYPICAL SIGNALS

Type of Signal	Analog Bandwidth Used (kilohertz)	Number of Bits per Sample	Digital Bit Rate Used or Needed (thousand bits/second)
Telephone voice	4	7	$7 = 56$
High-fidelity music	20	7	$20 \times 2 \times 7 = 280$
Picturephone	1000	3	$1000 \times 2 \times 3 = 6000$
Color television	4600	10	$4600 \times 2 \times 10 = 92,000$

human perception, plus the digital bit rate used or planned for their transmission with PCM.

In telephone transmission, the frequency range encoded in PCM is somewhat less than 200 to 3500 Hz. 8000 samples per second are used. Each sample is digitized using seven bits so that $2^7 = 128$ different volume levels can be distinguished. This gives $7 \times 8000 = 56,000$ bits per second. High-fidelity music with five times this frequency range would need five times as many bits per second.

In Picturephone encoding, a smaller number of bits are used to code each sample. It is not necessary to distinguish as many separate levels of brightness. The ratio between the bit rate and the bandwidth used is therefore smaller. On high-fidelity services such as network color TV, however, a larger number of bits is employed to minimize quantizing noise.

MULTIPLEXING   As noted, $4000 \times 2 \times 7 = 56,000$ bits per second are needed to carry a telephone voice. However, all the transmission facilities that this bit stream is likely to be sent over can carry a much higher bit rate than this. A thin pair of wires, such as those laid under city streets for ordinary telephone distribution, can be made to carry 1.5 million bits per second by using digital repeaters sufficiently closely spaced in manholes. Coaxial cables and microwave radio carry much more than this.

It is therefore worthwhile to send more than one telephone conversation over one pair of wires. This is done by interleaving the "samples" that are transmitted. If four voice signals are to be carried over one pair of wires, the samples are intermixed as follows:

> Sample from speech channel 1
> Sample from speech channel 2
> Sample from speech channel 3
> Sample from speech channel 4
> Sample from speech channel 1
> Sample from speech channel 2
> Sample from speech channel 3
> and so on

This is illustrated in Fig. 6.4. By sampling the signals at the appropriate instants in time, a train of PAM pulses is obtained; these pulses are then digitally encoded. For simplicity, only a four-bit code is shown in the diagram. Each PAM pulse is encoded as four bits. The result is a series of "frames," each of 16 bits. Each frame contains one sample of each signal.

In order to decode the signal, it is necessary to be sure where each "frame" begins. The signals can be reconstructed with this knowledge. The first four bits relate to speech channel 1, the second four to channel 2, and so on. A synchronization pattern must also be sent in order to know where each frame begins. This, in practice, can be done by the addition of one bit per frame. The added bits, when examined alone, form a unique bit pattern that must be recognized to establish the framing.

This process is called *time-division multiplexing*. It takes place at electronic speeds in computerlike logic circuits. The circuit components for digital multiplexing of this type are much lower in cost than those for frequency-division multiplexing. In the latter process, the range of frequencies available for transmission is divided up into smaller ranges, each of which carries one signal.

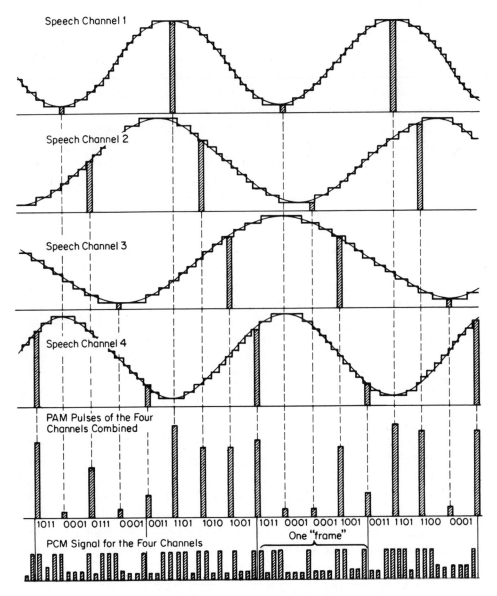

Figure 6.4. A simplified picture of time-division multi-plexing with PCM transmission.

THE BELL SYSTEM
T1 AND T2 CARRIERS

The most widely used transmission system at present with time-division multiplexing is the *Bell System T1 carrier*. This carrier uses wire pairs with digital repeaters spaced 6000 ft apart to carry approximately 1.5 million bits per second. Into this bit stream 24 speech channels are encoded, using pulse code modulation and time-division multiplexing. Eight thousand frames per second travel down the line, and each frame contains 24 samples of eight bits. Seven bits are the encoded sample; the eighth forms a bit stream for each speech channel, which contains network signaling and routing information.

The T1 carrier will be discussed in more detail in Chapter 17. Meanwhile, let us note that it is likely to be the basic building block of many of our future telecommunication networks. Other such systems will be built by other telecommunication companies and in other countries, but the ability to encode the basic telephone channel into 56,000 bits per second will be fundamental to them. The multiplexing of 24 such channels into a 1.5 megabit stream may become an international standard.

This technology is attractive for taking advantage of the vast quantities of wire-pair circuits that exist. The telephone companies have an enormous investment in these circuits, which span rural areas and are laid beneath the cities. The possibility now arises of making them carry 24-voice channels rather than the single channel that most of them now carry. Frequency-division multiplexing is also used on wire pairs, but this process normally gives 12-voice channels. Furthermore the cost of time-division multiplexing will increasingly tend to be less. Time-division multiplexing is also becoming attractive in the switching technology, as we shall see in Chapter 18.

When digital data are sent over a private analog voice channel, speeds of about 4800 bits per second are typical although 2400 is often used still. Twice this speed can be achieved with the penalty of a high-error rate. A rate of 2400 bits per second is used over a public channel with network signaling on it although here 1200 is still common. Some common carriers, particularly in countries other than the United States and Canada, permit transmission at speeds of only 1200, sometimes even less, on their public voice lines. These low speeds indicate the difficulty of designing *modulation* equipment to convert the data into a suitable form for traveling over the analog telephone line because of the high level of distortion on such lines.

The conversion to digital lines is good news for the data processing men. With synchronous transmission of seven-bit characters, the 56,000 bits per second of the PCM voice line give about 7500 characters per second, with a powerful error-detecting code. This is more than ten times the speeds in conventional use today.

Data processing specialists who have had reason to complain that the printer on the other end of a telephone line is slow, can be encouraged by the thought that over a PCM telephone line they could print at about 4000 lines per minute, or fill a very large screen full of data in a second. To do so, however, the high-capacity link would have to go into their premises. At present it commonly ends at the local telephone office.

The T1 carrier is only the beginning. A hierarchy of interlinking digital channels is planned for the Bell System. The next step up is the T2 carrier, which is designed to take the signals from four T1 carriers or, alternatively, to carry one Picturephone signal. This signal operates at 6.3 million bits per second. Higher still are the other T carriers which will carry hundreds of megabits per second over broadband transmission facilities.

The T1 carrier is already in operation on a large scale and is proving very successful. Production of the T2 carrier is starting at the time of writing. The higher T carriers are under test but not yet in production. As is discussed in Chapter 17, some problems in the use of the higher-speed channels are yet to be resolved.

One of the major needs of the 1970s is a public-switched network of channels for data transmission of widely varying speeds and flexibility. The T1 and T2 carriers are clearly candidates for the transmission links for this, although as we shall see there are alternative candidates.

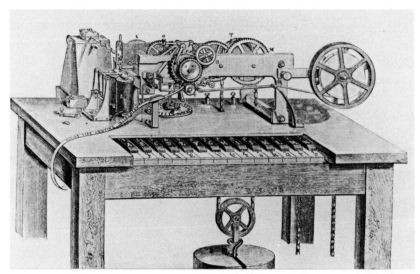

Hughes printing telegraph, designed 1855. Courtesy of The Post Office, England.

7 SPEEDS OF CHANNELS

In tomorrow's communication world, many different types of signals are going to be transmitted. By their nature they require widely differing transmission rates, or bandwidths.

Transmissions for human consumption require different rates. The eye can absorb about a thousand times as much information as the ear, when measured in terms of signal bandwidth. Similarly measured, high-fidelity stereophonic sound contains about ten times as much information as telephone sound. "High-fidelity" wall-screen television would require ten times the bandwidth of today's television—and today's television more than one hundred times that of high-fidelity sound.

When machines are considered, the variation could be even greater. Data transmission has primarily used a fairly restricted range of speeds so far. The most common speeds are the telegraph channel speeds, 45, 50, 75, and 150 bits per second; subvoice grade lines outside North America at 200 bits per second; and speeds derived from voice lines, 600, 1200, 2400, 4800, and now 9600 bits per second. Wideband data links operating at 19,200, 40,800, and 50,000 bits per second are becoming more widely used, and there is occasional use of higher speeds.

As mentioned earlier, analog and digital signals are interchangeable. We should keep this interchangeability in mind in discussing the speeds of future communication links. Computer data at 4800 bits per second can be converted into an analog form to travel over telephone channels of 3400 Hz. On the other hand, telephone signals can be digitized and represented by a stream of 56,000 bits per second.

Machines are in common use for sending data over voice channels at speeds of 600, 1200, 1800, 2400, 4800, 7200, and 9600 bits per second. Some machines are operating at speeds other than these. Speeds much

83

lower than 600 bits per second are also often used because of the convenience of attaching relatively slow machines to the ubiquitous public telephone network. It would be useful to standardize the speeds at which data are transmitted so that channels and machines become more easily interconnectable. It has been suggested that 600 N bits per second should be standard speeds, where N is an integer. Another suggestion has recommended 600×2^N, which gives less proliferation of equipment and facilitating multiplexing. Although standardization will eventually be of great importance, it can be restricting to enforce rigid standards too early in the development of a technology. The preceding standards would not fit conveniently into the Bell System hierarchy of PCM channels, the T1, T2, and T5 carriers, and these carriers, because of their widespread use, will almost certainly become standards at least within North America. (CCITT has standardized the sampling rate of 8000 samples per second.)

In the wide range of speeds at which transmission will take place, certain points stand out as being of particular importance. Some of these factors relate to human needs, such as the desire to make telephone sound intelligible and the caller recognizable with a low expenditure of bandwidth; some relate to machine needs, such as the maximum speed of mechanical printers; and some relate to channel properties, such as the maximum capacity of a wire pair.

Let us note the significant points in our range of speeds, as they indicate requirements for communication channels. Eventually it will be desirable to have all these channels available on a dial-up basis.

1. Very slow transmission

Some types of machines signal conditions to each other and no bulk of data is transmitted; for example, a remote burglar alarm transmits a simple Yes/No condition signal. A vehicle detector in the street transmits simple pulses as cars go over it. When you pick up or replace your telephone handset, this fact is transmitted to the relevant switching locations. Signals of these types require only a very small bandwidth.

2. Typewriter speeds

The input speed on a typewriter is the maximum speed at which a human being can type. Fifteen characters per second is enough for the most nimble-fingered. The output speed of a typewriterlike terminal need not necessarily be the same; however, this is today the speed of an economic printing mechanism. The price of a faster printer is substantially higher. There will be a massive requirement for channels of this speed, say 150 bits per second, in the foreseeable future. Today's telegraph channels of 75 bits per second and below are a little too slow for the fastest keyboard operators and for typical electrical typewriter mechanisms.

3. Human reading speed

If you are a fast reader, you might be able to read this page in a minute—in other words, at a speed of about 250 characters per second. If information is being displayed to you on a screen unit, it is desirable that it be transmitted as fast as you can read it. Doing so is a requirement of an efficient man-machine interface. Display screens operating today at 2400 bits per second (300 characters per second) on typical commercial applications do seem to provide an effective form of man-computer communication. Computer output that is substantially slower than this rate can be frustrating for the user. On computer-assisted instruction systems, it has been commented that a lower-speed output is like having a teacher with a speech impediment.

Sometimes a speed slightly higher than 300 characters per second seems desirable—for example, when the "page" is being skip-read, and the terminal user quickly flashes on the next page. This practice is likely to be common in browsing or searching operations. In looking through a telephone directory, for example, you do not read every line. When tables are displayed also, a fast operator can handle speeds somewhat higher than 300 characters per second. Many screen devices that can operate at 4800 bits per second (600 characters per second) are likely to be used today on analog voice lines. This rate seems to be a generally useful speed for such display terminals. It may well become the most commonly used speed on digital data networks.

It should be noted that although the terminal operator can usefully absorb 4800 bits per second, he certainly cannot respond at this speed (except by voice). A 150 bit-per-second return channel is adequate for his response with foreseeable mechanical devices, such as keyboards and light pens.

4. Telephone channels

The channel we are all most familiar with was designed with the important economic constraint that the bandwidth used should not be larger than necessary. The maximum number of telephone calls can then be multiplexed together over long-distance links or links between offices. The result is a bandwidth of about 3400 Hz, which is enough for you to recognize the voices of your callers and comprehend what they say. There is no need for high fidelity.

The systems we can design today are largely dominated by this bandwidth. It is highly desirable now to break away from this domination as quickly as possible.

5. Machine printing and reading speeds

A typical high-speed computer printer operates at about 1200 lines per minute—that is, about 20,000 bits per second. The highest speed card reader also reads cards at a speed close to 20,000 bits per second. It would be of value to have switched telecommunication lines interconnecting these machines with magnetic tape and disk units, and computers. Perhaps because of mechanical improvements, the input/output speeds will increase even more. Xerographic printing can give speeds that are higher than 20,000 bits/second. Optical document reading may also give higher speeds. However, they seem unlikely to require rates greater than two or three times the preceding speed. Standard broadband channels in increasing use in the United States and elsewhere operate at 40,800 and 50,000 bits per second.

6. A PCM voice channel

A voice channel using pulse code modulation transmits at about 56,000 bits per second. It is probable that many of the world's voice channels will use this form if transmission in the not-too-distant future. This rate is close to the speed requirement for high-speed printers and readers, and thus it would be valuable to have a switched channel of this speed.

7. A high-fidelity sound channel

Telephone channels are restricted to a bandwidth of 4000 Hz, of which about 3400 are usable. A bandwidth of 20,000 is needed for full high-fidelity transmission, and twice that amount is required if the transmission is stereophonic. The wires entering the home are capable of carrying this range of frequencies; therefore domestic distribution of high-fidelity music is technically feasible, and could be on a dial-up basis if desired. An analog channel engineered for high-fidelity transmission could carry about ten times as much information as a telephone channel.

If a pulse code modulation channel were designed for high-fidelity transmission, it is probable that at least twice as many detectable signal levels (quantizing levels) would be used as in speech transmission. The requirement would be eight bits per sample (explained in Chapter 17) instead of seven bits for speech. Perhaps even more bits would be used. If eight bits were used, the stereophonic transmission would be carried by about 640,000 bits per second instead of 56,000 for telephone speech.

Using a complex means of digitally encoding the analog signal (delta modulation—described later) a smaller number of bits can be used to carry music. This trade-off between bit rate and logical complexity of the digital encoding applies to most other types of digital transmission also.

8. The maximum capacity of a wire pair

Simple wire pairs are likely to be the most common method of carrying signals for decades to come. Multimillions of dollars worth of wire pairs are laid down under the streets and along the highways. In recent years the wire pair has been made to carry a bit rate of 1,500,000 bits per second or slightly higher. Digital repeaters at appropriately close intervals on the line, which regenerate, reshape, and retime the bits being transmitted, have been used. These repeaters are the basis of the Bell T1 carrier, which sends 1,544,000 bits per second over wire pairs with repeaters every 6000 ft. The T1 carrier is normally used to transmit 24 telephone conversations simultaneously. This is likely to be a standard for decades to come. A T1 carrier has been used for data transmission at 1,344,000 bits per second. Such speeds are appropriate for tape-to-tape or disk-to-disk transmission. This facility or a similar one seems likely to play a major role in data transmission because it makes the most effective use of the ubiquitous wire pairs.

9. Picturephone

When Bell System Picturephone signals are encoded in a digital form, 6 million bits per second are required. These bits can travel over a Bell T2 carrier. It is uncertain as yet how widespread the Picturephone service will become; however, such channels will probably become standard in the United States and will be used for many different types of transmission. Four T1 signals may be multiplexed together to travel over one T2 channel.

10. Television

The next step in speed relating to human communication is television, which requires about 4-6 MHz when transmitted over analog circuits. In North America one television transmission channel replaces two basic master groups that could have carried 1200 voice channels.[1] In the future, television, like every other type of signal, will probably be carried by digital pulse streams, and a channel operating at 92.5 million bits per second will be used for television and such a channel is planned in the Bell System future hierarchy of digital channels.

The television screen itself may eventually become larger and of higher fidelity. If the number of lines on the screen are doubled, then four times the bandwidth, or four times the number of bits per second, will be needed. If we have a 5-ft wall screen for television, or perhaps eventually a screen that occupies most of one wall, the resolution *per inch* will probably not need to be quite as good as that on today's small sets. Ten times the present bandwidth will probably be enough for even the most spectacular home screens. This would give approximately the resolution I obtain from my 35-mm slide projector, which has a 6-ft screen in my apartment and seems sharp enough for all practical purposes.

The ultimate requirement of visual transmission outside ultralarge-screen theaters may, then, be about 50 MHz or about one billion bits per second. Coaxial cables like those laid into homes today to carry CATV could be engineered to transmit such signals.

In all probability computers will find uses for this high transmission speed eventually. Signals to and from many terminals will be multiplexed onto one such channel. Responses to terminals on some systems will be in the form of pictures as well as alphanumeric responses. Data banks will be remote from the machines using data in them. Time-sharing systems of immense versatility will be able to call in programs from remote locations. Small machines will be able to handle highly elaborate applications with graphics by use of them on a fast dial-up basis.

11. The Bell T5 and T6 carriers

Today the upper levels of the Bell System planned hierarchy of PCM channels consists of the T5 and T6 carriers, which will transmit several hundred million bits per second. Such signals also may well become a standard for American or world digital channels.

[1]See Fig. 16.1 in *Telecommunications and the Computer.*

Figure 7.1   MCI's Proposed Carrier Network. MCI, Microwave Communications Inc., and its associated companies, have requested FCC permission to build a nationwide network of microwave links as shown. The links will be analog in operation and will use modems to provide digital channels. MCI plans to offer its customers leased transmission facilities, either analog or digital, customized to their needs.

72 basic channels will be available—ranging in bandwidth from 200 to 960,000 Hz. One-way or two-way channels can be leased and these can be split into subchannels as the user requires. MCI will provide digital channels from 75 bits per second up to high bit rates. There will be no restriction on how a channel is used. The customer can use it for telephone, data, facsimile, television, or any other traffic. He can attach terminal devices, modems, multiplexing and switching equipment. Voice and data channels can be alternated. Channels can be leased part time, and subscribers will be allowed to share channels. Subscribers can connect to the system at any relay station with any equipment they choose—for example, local telephone lines, direct microwave links, coaxial cable or mobile radio units. The customized nature of the service opens up new opportunities for equipment manufacturers.

The network shown links 165 cities in 41 states. As of early 1971, approval had been granted for the St. Louis-to-Chicago link only. The reader may compare this system with the Datran System described in Chapter 9.

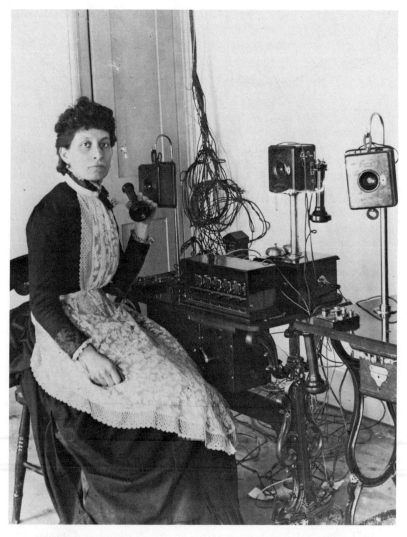

The telephone exchange at Croydon, London, 1884. Courtesy The Post Office, England.

8 SWITCHED DATA NETWORKS

In the world of intercommunicating machines that we envisage, one of the most important requirements will be a versatile dial-up network for the transmission of data.

In earlier decades it was a dream of the telephone companies that every telephone in a country would be able to dial every other telephone. Some even dreamed that all business offices and almost all houses would have such a telephone. In its day, this hope fired the imagination of the large numbers of people who made such a dream come true.

In today's dream we would like *machines* everywhere to be able to communicate with other machines. We would like to enter data into a terminal and dispatch the information at appropriate speeds to a data processing machine anywhere in the country or—and what is of rapidly increasing importance—anywhere in the world. We would like a burglar detector to send details to the police computer instantly. We would like machinery with an overheating bearing to dial for help. From our screen unit we would like to be able to browse in or query all kinds of computer files. Eventually, in a slightly more distant future, we would like to be able to set up a connection between our television set or wall screen and locations from which a picture can be received, cameras in friends apartments or in work locations, and videotape machines.

An immediate need is for a switched network for data transmission and several attempts are now being made to set up such a network. We can utilize today's telephone and telegraph networks to transmit data on a dial-up basis, and this step is being undertaken to an increasing degree.

91

There are, however, important requirements we would like the switched data network to meet, which today's networks do not. In many countries there is now discussion of a national computer "grid." A similar concept exists on a private basis within certain large corporations; and in countries in which competition is permitted in the building of telecommunication channels, such as the United States, attempts are being made by private enterprise to establish switched data facilities.

Ideally one would like to be able to enter data into a terminal and have the data transmitted at appropriate speed to a machine at any location. The addressing may be done by human dialing or by the machines themselves.

The requirements of such a system that differ from what is offered by present facilities are listed below.

1. Rate of transmission

The telephone operates at only one bandwidth. If data are sent efficiently over a dial-up telephone line, the speed is a few thousand bits per second, although lower speeds are often used. As we saw earlier, a wide variety of speeds ranging from very slow to exceedingly fast is needed. It is desirable that all these speeds of transmission eventually be available as switched public channels. Some corporations have installed their own switched data networks of high speeds, using leased lines to their own private switchgear.

2. Speed of connection

Manual dialing is slow; typically it takes about 20 seconds to dial a 10-digit number. It is faster with Touchtone dialing but still slow by computer standards. After the number has been dialed, there is a delay before the connection is established. The delay is sometimes a few seconds, often as many as 10 seconds, and may be as long as 30 seconds on long-distance calls. The delay is inherent in the electromechanical switchgear used for making the connection and for searching for alternate routes. These time delays are acceptable for human telephone conversation, but with switched connections for other purposes, we must reconsider what interconnection times are desirable.

For non-real-time data transmission, the end-to-end time can be quite long, as it is in some message-switching systems. For real-time work, a fast interconnection is needed, sometimes very fast. A request entered into a terminal needs a reply in a few seconds. Many airlines have specified that 90% of their transactions must have a response in 3 seconds or less, which requires a mean response time typically of about 1½ seconds. If such transactions, or some of them, were finding their way through a switched network, very fast switching would be required.

It is possible, as we shall see in Chapter 19, to build extremely fast switching equipment. Once fast-switched networks are in existence, a pattern of data processing that will emerge in many applications will be one in which real-time actions initiated at terminals sometimes require processing by more than one machine. When processing a transaction, a computer will request information or assistance from other computers. When booking a journey in which more than one airline (or other transportation medium) is involved, the computer handling the transaction will communicate in real time with the computers of other airlines. When booking a vacation, a travel agent's computer will interrogate many such sources. Data requested from a management information system or other form of information system will often require the interrogation of data banks in other locations.

In a switched data network planned by Datran for nationwide public use in the United States, the data path between any two subscribers will be established within 3 seconds. It is planned that this system should use a switch matrix of electromechanical components (reed relays).

Much faster switching is possible. One stored-program-controlled switch already in use has a nominal 10-millisecond connect time and a high-call handling capability. With this switch, end-to-end connection time could be a small fraction of a second, and multiple interconnections would be possible within the response time of a real-time terminal.

3. Automatic addressing

When a switched network is designed for machines, it seems natural that the "dialing" should be automatic. An addressing message may precede the transmission—to set up the appropriate path. Once it is set up, transmission proceeds. The addressing message would travel at the high data rate of the line, not at the slow speeds of today's dialing, or even today's multifrequency signaling between exchanges.

An alternative possibility is that the address needed for switching would travel with the message as a form of standard "header."

4. Cost

Use of today's analog channels for data transmission is expensive. The number of bits that can be sent over a voice channel is limited. This fact is particularly true for switched networks, on which the channel cannot be "conditioned" i.e. rendered sufficiently distortionless for fast rates of data transmission, and must carry network signaling with which the data must not interfere. In addition, the modems for converting the data into a suitable analog form and back again are costly.

Probably a network designed to carry only digital data could give a much lower cost per bit transmitted. Digital time-division-multiplexing techniques would be used which are becoming much less expensive than the high level of frequency-division multiplexing needed for analog circuits. The cost of the filters and modulation equipment for the combining together of analog signals is high. Prices proposed by the Datran Corporation in their filing for a digital network confirm that the drop in cost could be substantial. It is particularly so for transmission rates higher than those that can be handled by today's analog telephone circuits.

5. Speed conversion

Terminals connected to a switched data network will operate at many different speeds. Some method is required for accommodating the different speeds without undue wastage of channel capacity. The switched telephone network cannot handle higher-speed signals, and when low-speed signals are sent over it, as from the large numbers of typewriter-speed terminals thus connected, the available capacity is often less than 5% utilized. (Transmission at 15 characters per second over lines which can handle 300 is common practice in the data processing industry today.)

There are two ways of tackling this problem on a digital network. One is by means of time-division multiplexing in which the bits from the slow terminal are interleaved into bit streams from other devices so that full use can be made of channels between the multiplexing equipment. The other method is by means of digital storage built into the network and performing buffering operations. One attractive possibility is to have a slow device send data to a fast device and to have the speed conversion done in the digital network. We will discuss both time division multiplexing and network storage further below.

6. Broadcasting

Sometimes it is necessary to send a message from one location to several or many other locations. In a network designed for digital transmission, more than one address may precede the data, and the data will then be sent to all of them.

7. Code conversion

Different terminals use different transmission codes. An inexpensive terminal using, for example, five-bit Baudot code could be connected to a similarly inexpensive machine using seven-bit U.S. ASCII code, if the digital devices on the network performed the translation. The ability in a network of enabling incompatible terminals to communicate by automatic conversion of their character codes and line control procedures is referred to as "terminal transparency."

8. Data conversion or selection

It is sometimes advantageous to convert data as well as code for several reasons. They include privacy, error detection and correction, the compacting of information transmitted, the use of different language control messages in international transmission, and the structuring of responses that make the terminal easy to use.

It will be important in the years ahead to devise man-machine conversation structures that will enable people in all walks of life to communicate with the terminals. One can observe relatively few such structures today that make communication with the machines easy and natural, but one of the characteristics of those that do is well-worded explanatory items appearing on a screen at the appropriate moment. In conversation structures successful with the "man in the street," the machine usually says a lot and the man says relatively little. There is not necessarily any need to transmit the standard, verbose responses of the machine over long distances. They could be stored in devices near to the terminal and selected with a relatively short transmitted code.

If the terminals in question are inexpensive devices, perhaps in the home, the device for storage of responses will usually not be owned or leased by the subscriber. One possibility is that the generation of such responses could be done by the computer at the switching office that serves the subscriber. The responses may not always be printed words but may be in the form of pictures or spoken voice signals. Certainly in the *control* of switched networks, we are likely to find an increasing use of prerecorded voice responses to the users, responses generated by the network.

Some of the proponents of switched networks (not to mention regulatory bodies) say that the sole function of the network should be to transmit and switch. All data processing or conversion should be external to the network, which is merely a means of carrying information. They would exclude items 7 and 8 from this list. Item 7 does appear in some of the systems that are being considered for construction today.

SEPARATE OR INTEGRATED DATA NETWORK?   One of the controversial decisions that the telecommunication organizations have to make is whether the switched data network and the voice network should be integrated or not. Should telephone and data traffic travel over the same switched network? Should they share the trunks and the switching mechanisms or should they have entirely separate networks?

An extensive debate of the pros and cons has been taking place. The arguments changed entirely with the introduction of digital (PCM) circuits to carry telephone transmission. As we discussed in Chapter 6, a PCM channel designed to carry the human voice transmits 56,000 bits per second and could alternatively carry this rate of data. Many bit streams are interleaved (time-division multiplexing) between offices to give much higher-speed data channels. With suitable terminals, feeder channels, and switching, these bit streams could be used to carry all types of data. The switched data network and the telephone network could use the same trunks. On the other hand, the way switched telephone lines are used today for data is grossly inefficient. A long and expensive voice line might typically be obtained for half an hour to connect a teletype machine to a distant computer. In this time the user, working at human speed, with many pauses to think, might cause a thousand characters to be transmitted. With efficient organization the same line could transmit a million characters in the same time. If it is a PCM line, it could transmit ten million. If the same line formed part of a network designed for data transmission, this inefficiency could be avoided by interleaving many separate transmissions.

The advocates of a separate data network claim that in addition to greatly improved efficiency it would have many special features, such as those discussed above, which are not required in voice transmission. Furthermore, and this is a strong argument, the switched data network is needed *now* and the telephone authorities are not yet ready to make the desirable and extremely expensive changes in the telephone network.

In the United States, where there are many different common carriers, an organization not committed to voice transmission could build an entirely independent data network. One such attempt is described in the next chapter. Integration of the voice and data networks has no doubt been impeded by the Federal Communications Commission forcing AT & T into a corner labeled "telephone business only." In most other countries telecommunication facilities are under the control of the government, and thus one organization is responsible for both data and voice transmission. Some countries regard the data network requirement as merely an upgrading of their present Telex or other switched telegraph network. Studies on separate data networks are being conducted in many countries, and it seems likely that many such independent systems will be constructed. Germany is the first to have the system referred to as EDS (Electronic Data Switching System), which uses high-speed computer-controlled switching of data channels and gives the terminal users a choice of 1200 or 2400 bits per second transmission. For the machine uses we envisage, however, a much more versatile network is needed.

An attractive compromise in the question of whether or not the voice and data networks should be integrated is to construct a separate data network to meet today's needs but one designed in such a way that it can be integrated into a PCM voice telephone network at some future time. The telephone network in the United States is likely to be based on the Bell T1 carrier, and perhaps (one hopes) the digital format used in this carrier will become a standard in other countries as well. By the use of time-division multiplexing or appropriate buffering techniques, the data from terminals of widely differing speeds could pass through their own concentrators and switches and be linked to the PCM voice network.

CIRCUIT SWITCHING NETWORKS   The simplest form of network, and probably the one we shall see in the near future, is one that merely interconnects circuits between the terminal machines and that does not in any way modify or store the data sent. Channels of many different speeds may be provided. There are valid arguments, however, for building a network that stores the data and manipulates them in such a way that otherwise incompatible machines can communicate. The remainder of this chapter and the next will describe circuit switching networks, and Chapter 10 will discuss networks that store and manipulate the data.

Three system elements are necessary in order to provide end-to-end data communications:

1. The trunking system which provides the main highways between switching offices,
2. The switching system, and
3. The local distribution network.

Let us discuss these in turn.

THE TRUNKING
SYSTEM

Very high bit transmission rates can be obtained using microwave radio circuits, coaxial cables, or even wire pairs with sufficiently closely spaced repeaters. Such facilities will form the trunking system of future data networks. The transmission rate will generally be far too high for the users of the network; therefore the trunks must be made to carry the traffic of many users simultaneously. This process will probably be handled by interleaving the bit streams of different users through time-division multiplexing, in the same way that the Bell T1 carrier interleaves the bits of 24 digitized telephone signals.

In order to standardize and mass produce the channeling and terminal equipment, a basic operating speed may be selected for the main building blocks of the system. Channels of this speed will then be switched between the end offices. Many such channels will be multiplexed together over the high-capacity trunks.

The Datran System (discussed in the next chapter) proposes 4800 bits per second as its basic speed. Its trunking facilities are built from multiples of this speed, which is well suited to human reading capability and therefore apt to be the most commonly used speed for display terminals.

Other proposals have suggested that the PCM telephone channel should represent the standard speed and that a data network built around this speed would be capable of integration into the vast telephone network and into the Bell hierarchy of PCM channels, starting with the T1 carrier. As mentioned earlier, this carrier provides us with 8000 seven-bit "samples" per second, which we could alternately use for data. It is possible that one or two bits of each seven-bit sample will be used for synchronization and control purposes. If six bits of each sample are usable for data, we obtain a basic channel of $8000 \times 6 = 48,000$ bits per second, with the great advantage that it can travel over the world's future PCM telephone lines. It has been proposed that a data network be built in which channels of this speed be switched.

If a network is based on these speeds, signals from slower terminals will pass either through a submultiplexing stage before they enter the trunking system or through some form of "signal conditioning unit" that buffers them for transmission at the speed of the channel. Figure 8.1 illustrates some of the possible interconnections.

SWITCHING
SPEEDS
In the telephone network, we do not object to switching times of 10 seconds or more. In a data network, a much faster switching speed is advantageous for some applications.

Switching times in seconds may be adequate for the transmission of batches of data or items that do not need quick processing. However, many of the uses of the data network will be for real-time systems in which a fast response is needed. Several seconds of switching time for each transaction would be a major and often unacceptable handicap. Furthermore, a fast switching time is desirable for network efficiency. During the connection and disconnection process, a channel will be tied up for the duration of the switching time. Some of the messages on a real-time system are very short. Many systems designed for fluent and simple man-machine operation have terminal-user input messages of less than 10 characters. Ten characters take $16\frac{2}{3}$ milliseconds to transmit on a 4800 bit-per-second line. Between messages there will be a pause of human duration— many seconds.

Let us consider the timing for real-time conversational operation in a little more detail (assuming that its innumerable applications constitute the biggest eventual market for switched data networks). Timing considerations will depend on the basic operating speed of the channels.

We will consider two networks, one with a basic channel speed of 4800 bits per second and the other with a speed of 48,000 bits per second.

Let us take a fairly typical segment of human real-time terminal operation. It lasts 2 minutes, and in this time there are five conversional transactions, each consisting of an operator keyboard action, a response from the computer, and a pause while the operator thinks or composes his next input. It is desirable that the response from the computer and network be fast enough to facilitate a slick "conversation," and a response time in which the majority of transactions are answered in less than 3 seconds is a design criterion. We will assume that the time the computer takes to process a transaction and prepare its response is half a second.

First, consider a network that switches the connection in 3 seconds. If the above response-time criterion is adhered to, then the connection must be established at the start of the "conversation" and must remain connected for its duration (as would be the case with a conversation of this type on today's telephone network). The disadvantage is the extremely low efficiency with which the data links are utilized. If an average message and its response have 200 characters, the five transactions have 1000 characters, and the total time needed for transmission of these characters on the 4800 bit-per-second network is less than 200 milliseconds. Yet the channel is held for 2 minutes.

If the response-time requirement is relaxed somewhat, the connection might be reestablished for each new transaction, the "dialing" being done automatically by the terminal. The figures for different message lengths are then as shown in Table 8.1.

Table 8.1 MESSAGE LENGTHS FOR THE 4800 BITS-PER-SECOND CHANNEL*

Total Length of Inquiry and Response	Total Time for the Transmission of Data (milliseconds)	Total Time the Channel Is Occupied (seconds)	Transmission Efficiency (percent)
10 characters	16.67	3.517	0.4
50 characters	83.33	3.583	2.3
200 characters	333.3	3.833	8.7
1000 characters	1667	5.167	32

*The times for the transmission of data assume eight-bit characters.

For the 48,000 channels, the transmission efficiencies would be one tenth of the figures given.

It seems worthwhile that the switching time, especially for the faster channels, be more in keeping with the transmission times. It is certainly possible with today's technology to build a network with faster switching times than the above 3-second intervals.

Let us now assume an interconnect time of 200 milliseconds. With this switching speed, the line would probably not be held for the duration of the conversation; instead, the connection would be automatically reestablished for each transaction. Indeed, another question now arises. The computer takes half a second before the response is sent. Should the line be held during this time or not?

Table 8.2 THE TIMES FOR 4800 BITS-PER-SECOND TRANSMISSION*

Total Length of Inquiry and Response	Total Time for the Transmission of Data	If the line is held between input and response:			If the line is not held:		
		Total Transaction Time	Total Time the Channel Is Occupied	Transmission Efficiency (percent)	Total Transaction Time	Total Time the Channel Is Occupied	Transmission Efficiency (percent)
10 characters	16.67	717	717	2.3	917	417	4.0
50 characters	83.33	783	783	10.6	983	483	17.3
200 characters	333.3	1033	1033	32.2	1233	733	45.4
1000 characters	1667	2367	2367	70.4	2567	2067	80.6

*All times are in milliseconds.

Table 8.3 THE TIMES FOR THE 48,000 BITS-PER-SECOND TRANSMISSION

Total Length of Inquiry and Response	Total Time for the Transmission of Data	If the channel is held between input and response:			If the line is not held:		
		Total Transaction Time	Total Time the Channel Is Occupied	Transmission Efficiency (percent)	Total Transaction Time	Total Time the Channel Is Occupied	Transmission Efficiency (percent)
10 characters	1.667	701.7	701.7	0.24	901.7	401.7	0.41
50 characters	8.333	708.3	708.3	1.2	908.3	408.3	2.0
200 characters	33.33	733.3	733.3	4.5	933.3	433.3	7.7
1000 characters	166.7	866.7	866.7	19.2	1066.7	566.7	29.4

Here, especially with the shorter message, it becomes worthwhile to reestablish the connection between input and output.

The transmission efficiency with the PCM voice channel is still low for real-time transactions of typical length. Still faster switching speeds would be appropriate with this channel. Connection times as low as 20 milliseconds have been obtained with computer-controlled solid state switching, as will be discussed in Chapter 19. With a 20-millisecond interconnection time, the figures are as shown in Tables 8.4 and 8.5.

Thus it seems that very fast electronic switching is advantageous on a switched data network. If the network is to be geared to the speed of a PCM voice line for possible future integration with the telephone network, then switching speeds of tens of milliseconds seem appropriate. Such speeds are now practicable with computer-controlled semiconductor switch matrices with no moving parts. One machine that uses this switch technology is the IBM 2750 Voice and Data Exchange, which is marketed in Europe but (unfortunately) not in the United States (discussed in Chapter 19). The type of switching this machine uses would be ideal for a public data network.

LOCAL DISTRIBUTION NETWORK

There has been a great drop in cost in the long-distance trunks because of the higher bandwidth of microwave radio and coaxial cable. It is probable that solid state switching will reduce the cost of switching offices and increase their capability. The local loops from switching office to subscriber, however, have not dropped in cost, although multiplexers and concentrators using large-scale integration circuitry will enable many subscribers to use one such link (as in Fig. 8.1).

In general, the local loops are twisted wire pairs in multipair cable. Their loss-versus-frequency characteristics were designed for the requirement of telephone transmission only. It is desirable to use these wire pairs in the local data distribution network because of their wide availability and the enormous capital investment they represent.

There are two types of approaches to using the wire pairs for carrying a high bit rate. The first is to use equalization on the loop and design a modem that permits a wideband signal to be sent over it in an analog fashion. The second is to transmit a digital signal and to use repeaters at sufficiently frequent intervals on the line to regenerate a fast bit stream.

Taking the first approach, there are already examples in use of wideband transmission over local twisted wire-pair loops. The Bell System

Table 8.4 THE TIMES FOR THE 4800 BITS-PER-SECOND CHANNEL

Total Length of Inquiry and Response	Total Time for the Transmission of Data	If the channel is held between input and response:			If the channel is not held:		
		Total Transmission Time	Total Time the Channel Is Occupied	Transmission Efficiency (percent)	Total Transmission Time	Total Time the Channel Is Occupied	Transmission Efficiency (percent)
10 characters	16.67	537	537	3.1	556.7	56.7	29.4
50 characters	83.33	603	603	13.8	623.3	123.3	67.6
200 characters	333.3	853	853	39.1	873.3	373.3	89.2
1000 characters	1667	2187	2187	76.2	2207	1707	97.7

Table 8.5 THE TIMES FOR THE 48,000 BITS-PER-SECOND CHANNEL

Total Length of Inquiry and Response	Total Time for the Transmission of Data	If the channel is held between input and response:			If the channel is not held:		
		Total Transmission Time	Total Time the Channel Is Occupied	Transmission Efficiency (percent)	Total Transmission Time	Total Time the Channel Is Occupied	Transmission Efficiency (percent)
10 characters	1.667	521.7	521.7	0.32	541.7	41.7	4.0
50 characters	8.333	528.3	528.3	1.6	548.3	48.3	17.2
200 characters	33.33	553.3	553.3	6.0	573.3	73.3	45.5
1000 characters	166.7	686.7	686.7	24.3	706.7	206.7	80.6

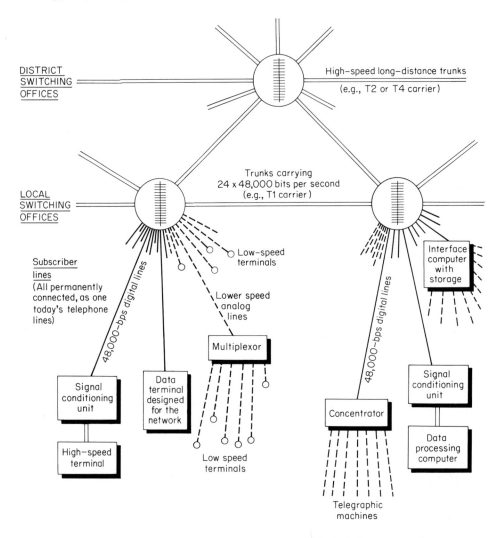

Figure 8.1. Interconnections to a switched data network based upon 48,000-bps channels compatible with PCM telephone channels.

303-type data sets, for example, transmit at as many as 230,400 bits per second over local loops and long-distance carrier circuits. Less complex and presumably less expensive modems could be produced which would carry 48,000 bits per second over wire pairs between, for example, the signal-conditioning units in Fig. 8.1 and the local switching offices. The attractiveness of this approach is that it requires no modification of the loops that now exist for telephones.

The second approach requires a repeater placing at intervals on the loop to regenerate the bit stream. Very high bit rates have been achieved over wire pairs in this way. On Bell T1 carrier circuits and elsewhere, transmission over wire pairs is being used at 1.5 million bits per second through digital repeaters at intervals of about 6000 ft. Less-frequent spacing is needed for 48,000 bits per second.

Figure 8.2 shows the attenuation in decibels per mile at different fre-

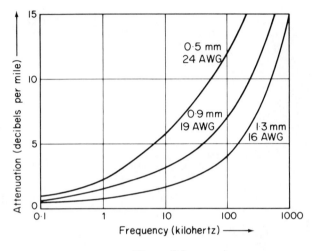

Figure 8.2

quencies on twisted-pair cables. The wire sizes for local loops generally vary from about 0.02 to 0.035 inch in diameter. Short-haul trunks are somewhat thicker, typically about 0.05 inch. Let us assume that the signal attenuation between the regenerative repeaters should not be more than 30 decibels. It will be seen that for the transmission of 1.5 million bits per second over 0.05 inch wires, a repeater spacing of little more than a mile is necessary. For transmission of our 48,000 bits-per-second signal over the local loops, a repeater spacing of 5 to 10 miles will probably be adequate. The loops themselves are frequently less than this distance in city areas. They average about 6000 feet in cities such as New York and Chicago.

Other forms of local distribution will probably emerge during the 1970s. The Datran System, described in the next chapter, uses low-intensity microwave links to bring the signal close to the subscriber and has installed experimental optical links. There have been various experiments with the use of signals of higher frequency than microwave for distribution in cities as we will discuss in Chapter 13.

A pony express rider in 1860 passing a telegraph line construction crew. The pony express went out of business shortly after the line was completed. Courtesy Western Union Telegraph Co.

9 THE DATRAN
    SYSTEM

In this chapter we shall discuss an example of a switched data network.

It seems an extraordinary way to set up in business: to start a corporation which will have no income for five years, and then spend more than a million dollars on market research. This is the story of the Datran Company. It reflects the high entry fee for anyone who wants to be in business offering switched telecommunications facilities.

Datran (Data Transmission Company) submitted a filing to the Federal Communications Commission[1] in 1969 proposing a nationwide network to carry data only. Its proposed charges for data transmission are much lower than existing tariffs. We shall make some comments contrasting the Datran approach with other proposals at the end of the next chapter.

The FCC's slowness in responding to the Datran proposal left Datran plenty of time to study their anticipated market. A highly detailed data bank of prospective users has been built, based on an extensive market study by Booz, Allen and Hamilton. The volume and type of transactions from every type of user was estimated, and this formed the input to a set of simulation programs which reveal both the performance of the system and the profitability of the company. Figure 9.1 shows a typical output of the programs used—a map of the United States giving the traffic densities at each location for one particular category of user. With the highly detailed model of the company and its system, the effect of all

---

[1]Sections in small print in this chapter are taken directly from the Datran filing. All figures are taken or redrawn from that filing.

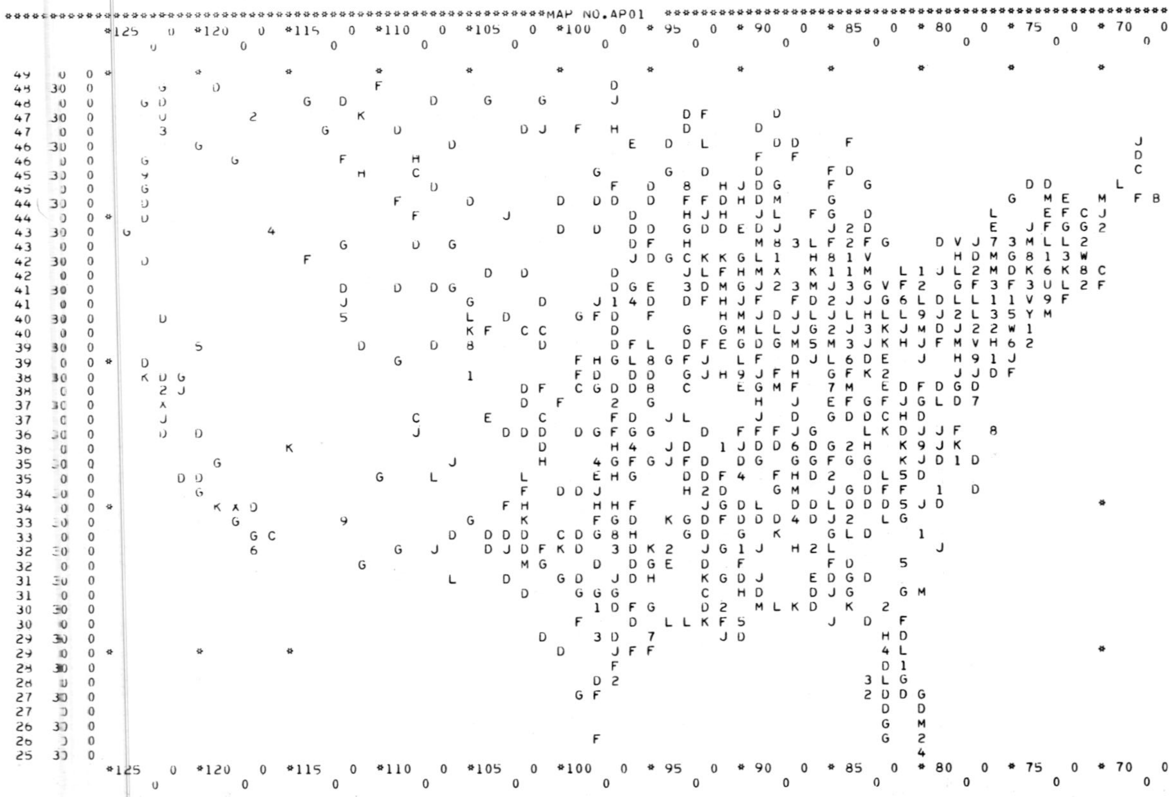

Figure 9.1. Map of density (Datran).
A map showing the forecast traffic density in the United States from Datran users of one particular industry. Each character on the map represents a specific load value. Such a map can be printed for any industry by means of a simulation program based on extensive market research. The simulation programs were used in designing the network and will help to give early warning of potential overloads when it is in operation.

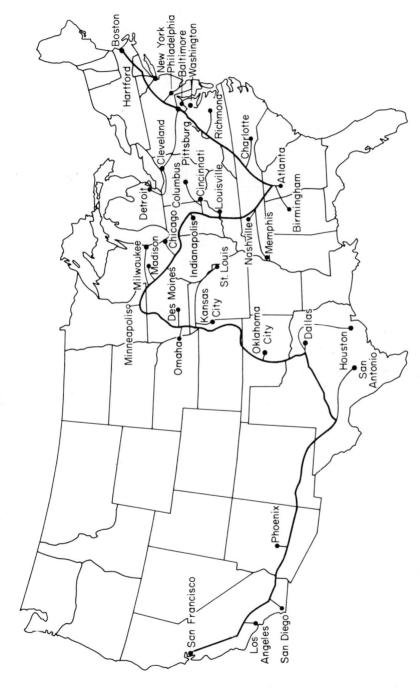

Figure 9.2. The backbone of the Datran system is a high-channel-density microwave trunk, as shown, using time-division multiplexing to provide an all-digital path.

109

possible changes in the market can be explored, and the changes in the engineering can be evaluated. Several major engineering decisions resulted from the use of the model. One member of the FCC is said to have commented that after a market study as thorough as this, if Datran does not succeed nobody would be likely to. The service to be provided fulfils a great need and the data processing industry can only lament that some such system was not "on-the-air" earlier.

Datran intends to build initially a high-channel-density microwave trunk, shown in Fig. 9.2, using time-division multiplexing to carry all digital transmission. Local distribution networks will be linked into this trunk, and data terminals operating at various speeds will be provided. A rapid switched connection between any two terminals will also be established. All switching will be controlled by computers. The network will be expanded according to market demands, possibly using satellites.

SYSTEM PHILOSOPHY   Datran describes its system philosophy as follows:

It is generally agreed that the market for data communications services will assume large proportions upon the availability of economical, switched digital services. The problem is how best to translate anticipated national market requirements into a realistic network design within constraints imposed by practical and regulatory considerations of financing, procurement and management.

The application recognizes that the demand for its services may not materialize precisely as initially forecast. Any forecast is necessarily a "snapshot" of a point in time, and the demand for data communication service will increase substantially and will vary in complexion in the years ahead. It is for this reason that in the design of the system great emphasis was placed on engineering flexibility. Channels of communication can be increased as needed to provide for an increase in traffic on a particular route.

The system switch and control is capable of optimizing the utilization of the transmission facilities by precise, instantaneous control of traffic routing. The applicant has determined that ten locations designated as District Offices and one location designated as a Regional Office will initially be needed to perform this function at the point in time of the market "snapshot" relied upon in this application. A modular technique has been adopted throughout the system to facilitate not only additions to the initial system capability, but rapid geographic augmentation to meet market demand.

The applicant's system is designed to provide interconnection capability with either TDM or analog modes of transmission. The applicant's policy will be to affirmatively seek interconnection with other carriers and other authorized communications entities in order to allow immediate service to all locations, and to augment the capability of the system to meet particular market service demands.

Satellite interconnection with the system is expected to be feasible although its implementation may depend on development of suitable terminal hardware to accommodate problems peculiar to the increased transmission distance of satellite transmission.

In addition to interconnection, it will be possible to integrate capabilities other than microwave into the system transmission. Cable applications will become more relevant in areas of high frequency congestion. Additionally, depending on the development and implementation of satellite communication capability, satellite TDM trunking may be possible on an integrated basis not only to provide an alternative to microwave in congested areas, but also to provide thin-route trunking capability.

In the design, particular stress was placed on system reliability and circuit availability. The system configuration is highly flexible and can be expanded easily. Conservation of the microwave frequency spectrum is important, and the time-division multiplexing of data utilizes the frequencies as efficiently as possible.

SERVICES TO BE OFFERED

Datran proposes to phase the services its system will provide. The initial services are described as follows:

1. Establish a switched point-to-point connection between two compatible subscribers within the network;
2. Manual or automatic addressing by the sender;
3. Abbreviated addressing;
4. Broadcast transmission to up to six compatible subscribers simultaneously;
5. Origination requested call back;
6. Controlled privacy.

The following additional service features will be offered if the market demand appears to justify them:

1. Speed conversion within specified ranges;
2. Code conversion between any two permissable code formats;
3. Speed and code conversion;
4. Expedited Information Transfer Service (EXIT) to provide the originating subscriber the option of forwarding data to a switching center with positive control over the time of delivery to the desired subscriber(s).

Transmission speeds

It is intended that transmission speeds of 150, 4800, 9600, and 14,400 bits per second will be offered initially, on a switched basis. Speeds of

19,200 and 48,000 bits per second will be available initially on a private service basis as market demand requires.

Charges

There are separate charges for "local" and "regional" transmission, but apart from this the charge is *independent of distance*. For other than very short distances, the charges proposed are much lower than common carriers charges at the time of writing for all transmission speeds, but especially so at the higher transmission speeds. The proposed charge for 14,400 bits per second is less than six times that for 150 bits per second, for example.

Connection time

A data path between any two subscribers will be established within 3 seconds of the completion of the addressing (on a keyboard like a Touchtone keyboard, or automatic machine addressing).

Error rate

The system is designed to give no more than one error in 10,000,000 bits transmitted, on average. This rate is substantially better than today's analog circuits, on which one bit error in 100,000 is more typical. Low-cost terminals with no error-detection or correction features would be more feasible in many data transmission operations than with today's circuits.

System availability

The system is designed to provide better than 99.98 % availability. In order words, a user's circuit will be out of operation less than 0.02 % of the time. This is a much better rate than is possible with today's leased channels.

Grade of service

"Busy" signals during the peak period should be encountered no more frequently than once in every one hundred attempts to obtain a circuit. In other words, the "grade of service" goal is 0.01. Except for the busy period, there should be very few busy signals.

SWITCHING   There will be two levels of switching office, the district office and the regional office (see Fig. 9.3). The district office provides the subscribers' connections, whereas the regional office maintains network control. Each regional office has direct control of up to ten district offices, and each district office can provide termination points for 1000 to 6000 terminals.

The salient functions to be performed by each office are as follows:

District Office

1. Provides subscriber terminations
2. Responds to all requests for service
3. Insures subscriber-to-subscriber compatibility via class code distinction
4. Determines and establishes intraoffice switch linkage
5. Coordinates with Regional Office trunk assignments for interoffice transmission
6. Maintains records of all services provided to each subscriber (for billing purposes)
7. Maintains necessary statistical information for future analysis
8. Provides maintenance status and suspect component identification

Regional Office

1. Maintains a complete network directory
2. Assigns all trunks within its area of jurisdiction
3. Determines and establishes intraoffice switch linkage
4. Establishes alternate path as required
5. Collects network use information from each District Office at prescribed intervals
6. Maintains necessary statistical information for future analysis
7. Provides maintenance status and suspect component identification

Each office is controlled by a computer, which is engineered for reliability; in case of failure, a standby computer takes over. The processor scans the incoming lines for a request for service, receiving and interpreting the address sent and connecting a path through the switch matrix appropriately. At the end of the call, it disconnects the switches and records information about the call for billing and statistical analysis. Computer-controlled switching will be discussed in Chapter 19.

TRUNKING   The high-density backbone of the system, shown in Fig. 9.2, uses microwave radio transmission between antennas within line of sight of each other on appropriately selected high locations. It is routed so as to serve the major data

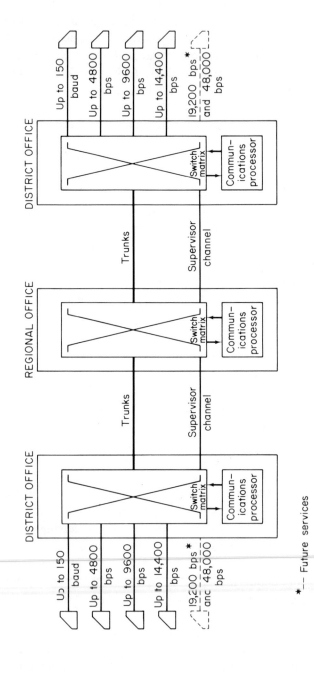

Figure 9.3. The Datran System switching offices. In both the district and regional offices the switch matrix is operated at high speed under computer control.

★-- Future services

114

concentration areas of the United States. Spur trunks also using microwave and identical electronics carry data to the district offices shown which are off the backbone trunk route.

At each microwave station on the backbone trunks and its spurs, the digital signal is relayed by a repeater that receives, amplifies, and reconstructs the bit pattern. It transmits a new, clean, conditioned, and retimed signal. Noise is therefore not cumulative as it is on analog systems. This is one factor that enables the system to have an error rate on the order of 100 times lower than on today's analog circuits.

Some of the microwave stations have a simple repeater; others have a "branching repeater," which the additional function of inserting or extracting some of the channels. The channels may be terminated at that point and taken into a switching office or they can be transmitted over a spur to locations not on the backbone route.

Figure 9.4 shows the branching repeaters and the proposed initial capacities of the trunks in terms of 4800 bits-per-second channels. Considerable expansion of these channel volumes is possible.

It is planned that each trunking station shall have alarm and control facilities. The station will have 32 alarm functions to alert the supervisory locations to any malfunctioning of the network. The status of the network and the remote sites will be automatically monitored 24 hours per day. In addition, each station will have 16 control functions for controlling the network. The alarm and control signals will travel along a data channel in each direction, submultiplexed for this purpose.

The channels that are specially reserved for control and signaling are referred to as "order wire" channels (in memory of earlier technology). In addition to the channel for alarm and control, a channel will be used to permit station-to-station voice conversion for control purposes. The voice will be digitized with a sampling rate of about 20,000 bits per second.

MULTIPLEXING   Being an entirely digital network, time-division multiplexing is used throughout, and this enables the maximum quantity of data to be sent over the microwave links. In all parts of the network, the bits are transmitted by using simple phase-shift keying (i.e., information is encoded by changes in phase).[2]

The basic channel of the system transmits 4800 bits per second. In order to transmit 9600 bits per second, two of these are used; for 14,400, three are used. Higher multiples can permit higher data rates from a terminal. Thirty-two low-speed channels operating at 150 bits per sec-

2See *Telecommunications and the Computer,* Chap. 13.

ond are derived from the 4800 bits-per-second channel, again by time-division multiplexing.

The multiplexing equipment is modular in design so that it can be installed economically and different terminal speeds can be accommo-

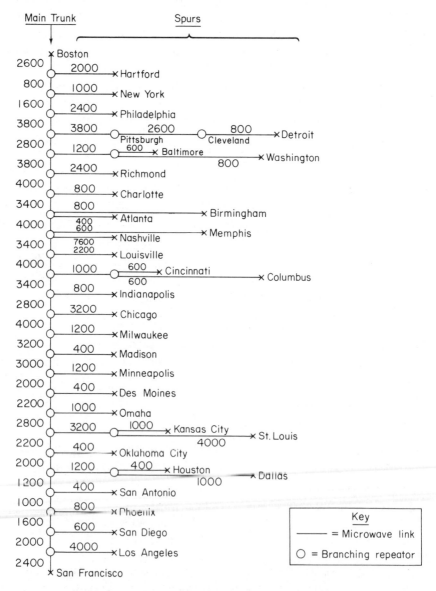

Figure 9.4. The figures indicate the number of 4800-bps data channels in the original FCC filing for the Datran System. These are channels in the microwave trunk shown in Fig. 9.2.

dated easily. Additional channels can be provided on a plug-in basis. The change from 4800 to 9600 bits per second or higher requires only a wiring change.

The microwave radio path, as initially planned, can carry up to about 4000 of the 4800-bits-per-second channels. This figure can be increased when the market demands it. A "frame" of bits for transmission consists of one bit from each channel, plus one timing bit. This bit stream (shown in Fig. 9.5) is used to modulate the microwave radio frequency transmitted. At a branching repeater station, the signal is demodulated, separated in 4800 bit per second channels, and these are switched as required. The resulting groups of channels are then remultiplexed and transmitted.

LOCAL DISTRIBUTION

It is important to construct a local distribution system with as high a quality as the long-distance transmission. The local circuits in conventional use today for data transmission give an error rate that would substantially degrade the end-to-end performance of this system.

In the Datran filing, the continuity of the digital signal is maintained from its point of origination to its destination. No digital-to-analog conversion is required. The advantage of being able to reconstruct the bit stream at repeaters is maintained to its termination.

It is proposed that distribution in the cities be via low-powered microwave equipment. A band relatively free of frequency congestion, the 11 GHz common carrier band, will be used. Subscribers in a typical city tend to be clustered. Thus there will be concentration points of relatively high density, such as business areas, industrial parks, and large office buildings. It is economical to take the microwave links to these groups. Less-dense areas will be microwave connected if they provide sufficient business to justify this cost. The low-density microwave links will connect such areas to the district office (Fig. 9.6). Shielded multipair cables will connect other nearby subscriber locations to the microwave antennas.

Approximately 50 such microwave stations can link to the district office; these may be connected in a multitier or ring configuration. The radio links have a maximum distance of 5 miles—much shorter than the spacing on the trunk routes.

The Datran Corporation selected a typical large city and applied the principles of its local distribution network in detail. Potential customers were identified; the locations of high-rise buildings were noted; banks, schools, hospitals, government buildings, industrial complexes, and so on were analyzed to develop cluster areas for siting the microwave an-

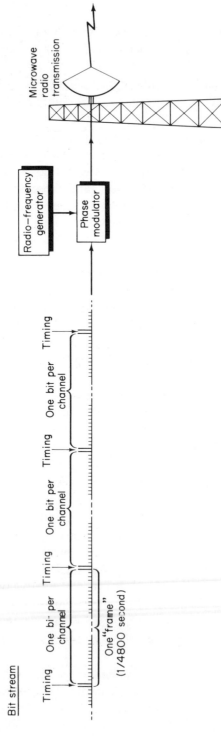

Figure 9.5. Initially, 4000 channels of 4800 bits per second each will be multiplexed together for transmission over the microwave system.

118

Figure 9.6. Datran's local distribution system will use low-power 11-GHz microwave transmission, and short laser or infrared links. A local "drop point" (right) receives the signal, in this diagram, by microwave, and relays it to subscribers by underground cable or roof-to-roof transmission *Courtesy Data Transmission Company, Falls Church, Virginia.*

tennas. Figures 9.7 and 9.8 show typical configurations that resulted. The configuration in Fig. 9.8 could service about 1700 subscriber terminals operating at 4800 bits per second.

The microwave equipment will be roof-mounted in cabinets that are 8 cu ft in size, with a 4-ft parabolic reflector for the radio antenna. The antenna will be mounted on the building in a manner to provide shielding to minimize mutual interference with other stations. The low-power levels used in the transmitters relieve this problem.

In addition to microwave channels, optical transmission between city rooftops is being used experimentally. The experiments appear to be successful and some of the city distribution may eventually be done with beams of light, or infrared transmission. Such links, however, can be put out of action by very heavy storms or thick fog. This is discussed in a later chapter.

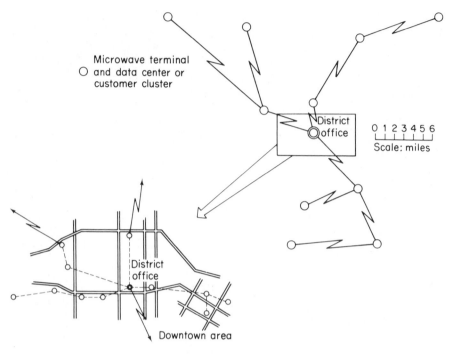

Fig. 9.7.   Low-power microwave links in a city center.

This system is highly flexible and can accommodate subscriber change easily.

CONNECTION VIA EXISTING ANALOG CIRCUITS

Not every location will be served by the digital network. Those locations beyond its reach may be connected to it by existing communication links. The Datran filing says:

> If requested to furnish service to subscribers located outside of tariffed areas, the applicant is prepared to enter into contractual arrangements with other common carriers serving such areas. To the subscriber who requests use of existing common carrier local distribution facilities, the applicant will make available modems and the Digital Communications Console to permit such interconnection. However, in these circumstances, the subscriber's data transfer rate will be limited by the class of service by the interconnecting common carrier.

The old-established common carriers bitterly refer to this as "cream-skimming," saying that Datron wants to operate on the high-density, high-profit routes only, whereas they are compelled to give universal coverage. We shall discuss this argument in Chapter 23. It might make sense for the existing carriers to be permitted to "cream-skim" also if this provides the data processing community with badly needed facilities.

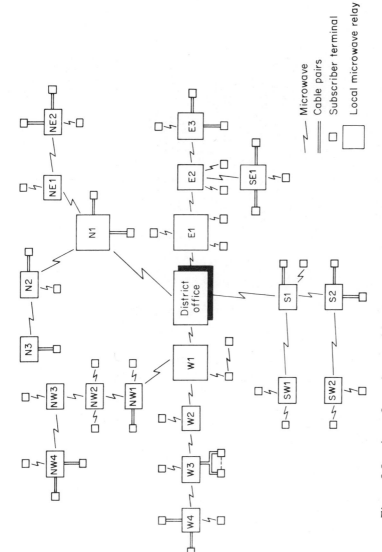

Figure 9.8. A configuration resulting from a study of a typical large city in the United States, showing the use of low-density microwave connections and cable pairs to the subscriber terminal locations. *Reproduced from Datran FCC filing.*

An advertisement of the Great Western Railway, England, 1845. Courtesy The Post Office, England.

10 A "HOT POTATO" NETWORK

The last two chapters described data networks in which physical circuits are switched, like telephone lines. This chapter discusses a fundamentally different approach to constructing a data network—one with some particularly attractive features.

CIRCUIT SWITCHING
VERSUS MESSAGE
SWITCHING

There are two ways in which data can be switched so that they are transmitted to a network destination, circuit switching and message switching.

With *circuit switching*, the switch mechanism interconnects a physical path between the parties that transmit to each other. The signal from one telephone passes directly and almost instantly to the telephone it has established connection with. The signal may undergo a variety of electronic contortions, but it is transmitted as though there were a direct metal connection between the telephones; for short interconnections, there is. The switch mechanisms set up this physical path.

The communication of data and telegraphic information has long since utilized *storage* in a way that is not possible for voice or television. This factor is the basis of *message switching*. The data are sent to a location where they are stored. At a convenient time the information is taken from the store and transmitted to its destination or to the next relay point. Two operations occur simultaneously in a message-switching location— the receiving and storing of messages to be relayed and the removing of them from storage and transmitting them. There is no physical path between the line on which the message arrives and that on which it is sent out. The message will be delayed at the relay point. At a manually operated message-switching station, the delay could be minutes or hours. With a suitably designed computer, plus storage that is accessible in microseconds, the delay could be very short.[1]

Store-and-forward systems of this type permit broadcasting to many recipients; they allow incompatible terminals to send data to one another and allow data to be stored until a terminal that is occupied or unattended becomes available for use. High line utilization can be achieved with such systems. Many such systems have been used as adjuncts to data processing systems.

"PACKETS" OF DATA
It has been advocated, originally by P. Baron of the RAND Corporation and later by D. W. Davies and his co-workers at the British National Physical Laboratory (see references at end of chapter), that a network of very fast *message switching* computers which route a specially formatted package of data onwards at high speed could provide a more flexible data network than could be built using circuit switching.

Data from a subscriber, in any form, would be placed by a local network computer in a data "envelope" with a header containing the destination address and other control information. This envelope is referred to as a *packet*. The packets are transmitted on very high speed links between the nodes of a nationwide network. Each node is a small computer that passes the packet on to the next point in the network, like a hot potato, as fast as it can. The time taken at one node to relay the packet is *only a few milliseconds*. On reaching their destination, the data would be removed from the packet by the local computer and sent to the terminal to which they were addressed.

There are many possible variations on this theme. In Davies' system the packets are of various sizes, being made of one to eight segments of 16 bytes (of eight bits); the maximum is 128 bytes. The first segment contains the address and control information. A long data message will require many such packets, but the majority of messages on real-time systems would fit into one packet.

Davies proposed a public switched network. The first network to come into operation using these principles was a private one, the ARPA network (U.S. Government Advanced Research Projects Agency) linking dissimilar computers at ten ARPA-supported research centers at MIT,

[1]Message-switching systems are described in Chapter 17 of the author's *Design of Real-Time Computer Systems*.

UCLA, the RAND Corporation, the Stanford Research Institute and other such locations.

THE INTERFACE
COMPUTER

An important principle is that the formatting of the packets is the responsibility of the network, not of the user. The formatting is done by an *interface computer*. A configuration of such computers in Davies' proposal is shown in Fig. 10.1. The line from the interface computer to the terminal is permanently assigned to the subscriber as is the line from the local central office (telephone exchange) to a subscriber's telephone set. The interface computer is likely, in fact, to be located at the local central office in many instances.

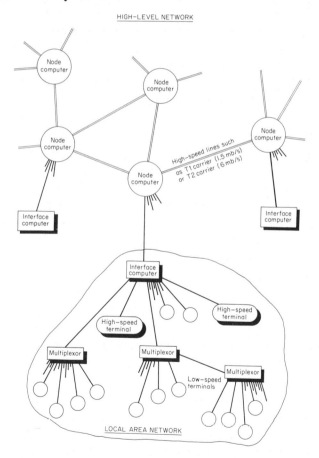

Figure 10.1. Adapted from "Principles of a Data Communication Network for Computers and Remote Peripherals" by D. W. Davies. IFIP Congress, Edinburgh (1968).

computer is likely, in fact, to be located at the local central office in many instances.

A wide variety of different terminals can be attached to the interface computer. Indeed, one of the chief advantages of this type of network is that entirely different types of terminals can intercommunicate. They can have widely different speeds and use different codes. They can be synchronous or start-stop. They can use polling, contention, or be alone on a line. They can use different types of error control. If desired, they can be connected via multiplexers or concentrators. Above all, they can be inexpensive, for most of the costly terminal features like buffering and elaborate line control are not really needed. The interface computer maintains a list of the characteristics of all the terminals attached to it, their control mechanisms, speeds, and transmission codes. If the code differs from the network transmission code, the interface computer converts it as the packet is being assembled. When the packet is received by the destination line control computer, it is converted, if necessary, to the code of the receiving terminal. It is estimated that one interface computer could handle more than a thousand terminals in this way. It is possible that some data processing machines will do their own interfacing and assemble their packets themselves.

An additional function of the interface computer is to collect the information necessary for logging and billing subscribers.

THE HIGH-LEVEL NETWORK   The interface computer sends its packet to the nearest node computer, and the packet quickly passes through the high-level network to the interface computer at its destination. The full-duplex links between the node computers can be of differing speeds. A typical speed in the future may be 1.5 million bits per second (the speed of the Bell T1 carrier). At this speed, a maximum-size packet will take 0.68 millisecond to be transmitted between nodes. At the speed of the T2 carrier, it would take 0.17 millisecond (not including the propagation time.)

The node computer receiving the packet places it in an input queue. When it reaches the head of the queue, the packet is placed in an output queue for the next node computer most suitable for it at that instant. It will be sent to the computer closest to its destination provided that (1) the computer and transmission line are operating correctly, (2) the output queue for transmission to that machine is not full, and (3) the input queue of that machine is not full. When an input queue becomes full, the computer holding it instructs its neighbours not to send any more until the congestion abates.

The packet thus zips through the network, finding the best way to go at each node and avoiding congested and faulty nodes. The mean total delay of a packet passing through a node is typically about 1 millisecond. The end-to-end journey of a packet traversing the country might normally pass through about five nodes. The response time of the network is therefore very fast.

Programs for the main functions in the node computer were written at the National Physical Laboratory. A small 16-bit word machine with a cycle time of 0.6 microsecond was used. With a mean packet length of seven segments (112 bytes), it was found that 2600 packets could be relayed per second.

The cost per packet and the performance of the system are of a quite different order to that conventionally associated with message switching. The reason is that the packet format is designed for computer handling and its contents need no processing in the node machine. Another reason is the speed of the lines. The time for transmitting and receiving the packets is of the same order as that for processing them.

It is possible that packets in this network could fail to reach their destination because of a temporary equipment failure or data error in the address. Such packets might be passed indefinitely from one node to another if something did not stop them. To prevent this occurrence, a *count* field is used in each packet and the number of nodes that have relayed that packet is recorded in it. When the count exceeds a certain number, the packet is returned to its point of origin. This process protects the network from becoming clogged with roving, undeliverable messages.

NETWORK AND ERROR CONTROL   The format of a packet carrying a message is shown in Fig. 10.2. It will be seen that an eight-segment packet can carry 112 bytes of data.

Each segment has seven bits for the detection of errors in that segment. These bits are checked when the packet is received at each node. If an error is found, retransmission of that packet is requested. The packet is therefore not deleted in a node when it is transmitted, but instead is placed in a "trace queue." It can be deleted from the trace queue either when a given amount of time has elapsed, so that it is certain that retransmission will not be requested, or when a control message is received.

Control messages have the same format as a one-segment data packet and are transmitted in the same way. Their first four bits indicate that they are a control message. The following types of control message are used:

4 bits indicating the type of packet
 12 bits giving its identification
  20 bits giving the destination address
   20 bits giving the origination address
    The data being sent
     A count of the number of nodes this packet
     has passed through
      One bit indicating whether there
      is a continuing segment

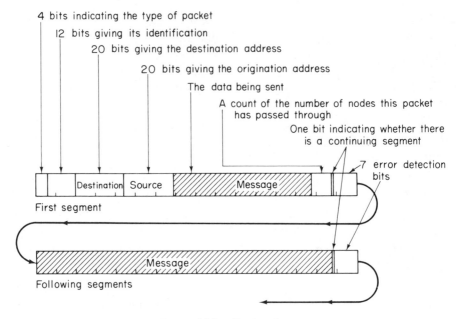

Figure 10.2. Packet format.

1. *Trace*—to cause retransmission of a packet that has become damaged. The trace message contains the identification number of the packet (the next 12 bits).
2. *Shut-up* tells a neighboring node not to send any more messages because the input queue is full.
3. *Breakdown* is generated when a failure in a node or link is detected.
4. *Start* tells a node after a *shut-up* or *breakdown* condition that transmission can begin again.

The error control that takes place between the computers in the network is independent of error control between the interface computer and the terminal. The latter will depend on the design of the terminal and the programming for control on the terminal line.

PRIVATE LINES   The way the terminals can be used depends on the organization in the interface computer. In many data processing systems, a permanent connection is needed between a terminal dedicated to a particular job and its associated computer. In this case, the interface computer to which such terminals are attached would store the permanent interconnection address. The terminal need never address a message. Any data from it would be sent automatically to the same location. This process is equivalent to having a private line from that terminal.

Similarly, the destination address could be stored by the interface computer for the duration of a terminal "conversation."

CONTINUOUS
TRANSMISSION

Continuous transmission of data could take place in the system described at speeds equivalent to today's broadband tariffs.

Where transmission originates at a computer or at a concentrator or control unit with sufficient logic capability, the functions of the interface computer could be bypassed. The originating machine would format its own packets. The packet buffering would be in the originating (or receiving) machine rather than in the interface computer. It is suggested that this mode of transmission, referred to as "cut-through," might be offered at a lower price.

THE ARPA
NETWORK

Figure 10.3 shows the ARPA network that is based on countrywide 50-K bps links. The interface computer does not handle terminals in this network, and is, itself, the switching node. Many new facilities are likely to be added to the network shown. The network is operating successfully and the round-trip transmission time is about 100 milliseconds.

ECONOMICS

Whether circuit switching or message switching will ultimately give the lowest-cost network for the distribution of data remains to be seen. The British National Physical Laboratory made rough estimates of the cost of such a system based on the assumption that 10,000 terminals were in use in the United Kingdom and that the high-level network for these terminals had 20 nodes. The costs for the high-level network looked very low: 0.1 ¢ per packet transmitted (independent of distance). Most of the cost lay in the local distribution network. Based on 1968 technology, these costs worked out at $2400 per terminal for the wire pairs, concentrators, and logic required at the ends and at about $240 to $480 per subscriber for the interface computer. These are capital costs on equipment that would have a long life; they would be lower with such emerging techniques as large-scale integration.

The full cost of alternatives would include modems and, at some terminals, buffers and control logic, which are dispensed with here because the interface computer does the buffering.

ADVANTAGES AND
DISADVANTAGES

The advantages of this type of system derive from two different aspects. It is worthwhile to differentiate between them. The first is the "hot potato" aspect of the high-level network, and the second is the storage and message manipulation afforded by the interface computer. Either aspect could be retained in some form without the other.

The main advantage of the "hot potato" routing is that trouble spots are avoided. No matter where they are, the packet finds its way around them unless the destination computer itself is faulty. The interconnection is likely to be more reliable than the fixed private line on many of today's real-time systems. When one part of the network becomes congested, again, the packet will find its way around this spot. In any circuit-switching network, the subscriber will occasionally receive a "busy" signal. The Datran System designers estimated that this would happen one percent of the time during the peak periods. In the packet-passing network, the subscriber need never receive a busy signal. During the momentary periods of congestion, packets are queued until they can be sent, and the queuing space in the interface computers can be made large. When the entire network is congested, preference can be given to high-priority packets, such as those from real-time systems. The packets for batch processing will be put aside until there is capacity to send them. Generally the wait will only be a small fraction of a second.

The overwhelming advantage of the interface computer is that it allows machines of all different types to communicate. It gives the network great flexibility, for all kinds of new devices can be connected to it when they are invented. If, for example, a new type of graph-plotting device should be invented 5 years from now, it is probable that there will be no problem in connecting it to this network.

The use of storage in the network has a number of advantages. Even if circuit switching rather than message switching is eventually used, it would be worthwhile to employ message storage at certain points in the network.

The possible advantages of the use of message storage in a network are given below.

1. Buffering

It is desirable that many of the terminals connected to the network be as inexpensive as possible. They should be unbuffered and asynchronous. Asynchronous characters arriving at the local switching center, as an operator pounds her keyboard, can be stored in a time-shared buffer and then transmitted at the appropriate circuit speed.

2. Speed conversion

By using such buffering, terminals operating at different speeds can be interconnected.

3. Broadcasting

The sending of one message to many addresses can be facilitated. (This factor is also possible using computer-controlled circuit switching.)

4. Message holding

Data for a terminal that is busy or otherwise unable to receive can be stored until the terminal becomes free.

5. Message filing

A service could be offered in which certain messages are filed for a period of time and then referred to on demand.

6. Automatic handling of transmission errors

The network could be designed to attach its own error-detecting characters to the messages and retransmit them if an error is found.

7. A nonblocking network can be constructed

With simple circuit switching, some attempted interconnections would receive a busy signal either because the recipient was receiving another call or because the necessary lines or switch paths were occupied. In a network with storage, the message could be held until the required path became free and the sender would never receive a busy signal.

STORAGE IN THE NETWORK Storage could be employed in a data network in many different ways. It could be used in a system that is basically a circuit-switched rather than a message-switched network. Indeed, some form of storage will often be desirable as a buffer between terminal speed and line speed.

Basically, storage can be employed in three places. First, it can be at the terminal, for buffering purposes—not a part of the network at all. Second, storage could be at the local switching office. It is cheaper to have time-shared buffering there than at each terminal. Third, storage could be at a higher-level switching office, perhaps with the local distribution portion of the network going through circuit switches and the high-level portion being a packet-passing network as described.

The facility of the interface computer to permit connections between incompatible machines is one of the most advantageous features of the National Physical Laboratory system, and this factor could be retained on a system that otherwise uses circuit switching.

THE BEST OF ALL WORLDS

In the switched data networks described in this and the previous two chapters, there are a number of excellent technological ideas. The best ideas, however, are not all found on the same system.

The Datran System would be better if it had high-speed switching, as in the IBM 2750. The National Physical Laboratory System might have lower local distribution costs if it used microwave or optical links in the cities, as the Datran System does. Either system could be integrated into the Bell T1 and T2 carrier networks so that voice and data use the same trunks although not the same switching. The National Physical Laboratory interface computer enabling interconnections between entirely different types of terminals has great appeal.

A network combining the best features might have the following aspects:

1. Use of the Bell T1, T2, and higher-carrier systems, or compatible systems, for trunking.
2. Use of satellites for long-distance interconnections.
3. Microwave or optical local distribution in the cities, like the Datran System.
4. Switching in milliseconds as made possible by solid-state switching (like the IBM 2750) with alternate routing for reliability.
5. Fast "packet" switching where this factor gives lower costs than circuit switching.
6. Subscriber interface computers so that all kinds of machines can intercommunicate and inexpensive unbuffered terminals can be used.

DANGERS OF INCOMPATIBLE SYSTEMS

There seems to be great danger at present that commitments will be made in many countries to systems that are only half adequate and that cannot be developed easily into networks truly meeting the vital needs of the computer industry.

A further danger, and one that should be taken very seriously is the danger that data networks installed in different nations will be incompatible. The networks favored in different countries at the time of writing would require different terminal equipment and different software. Interconnection of the networks would be expensive. Never before has there been such an acute need for international cooperation and standardization in telecommunications.

REFERENCES

1. P. Baran, "On Distributed Communication," RAND Corporation Memorandum, RM-3420-PR, August 1964.
2. D. W. Davies, "The Principles of a Data Communication Network for Computers and Remote Peripherals," presented at IFIP Congress, Edinburgh, August 1968.
3. R. A. Scantlebury, P. T. Wilkinson, and K. A. Bartlett, "The Design of a Message-Switching Centre for a Digital Communication Network," presented at IFIP Congress, Edinburgh, August 1968.
4. K. A. Bartlett, "Transmission Control in a Local Data Network," presented at IFIP Congress, Edinburgh, August 1968.
5. L. Roberts, "Computer Network Development to Achieve Resource Sharing." AFIPS Spring Joint Computer Conference.
6. F. Heart, R. Kahn, S. Ornstein, W. Crowther, D. Walden, "The Interface Message Processor for the ARPA Computer Network." AFIPS Spring Joint Computer Conference.
7. H. Frank, I. Frisch, W. Chou, "Topological Considerations in the Design of the ARPA Computer Network." AFIPS Spring Joint Computer Conference.

Scaffolding for wires coming into
telephone exchanges, London, 1909.
Courtesy The Post Office, England.

11 THE TOUCHTONE TELEPHONE AND VOICE ANSWERBACK

The cheapest way to converse with a remote computer today is by means of a conventional telephone. The telephone first dials the computer, and when the connection is established, the same instrument is used for sending data. This operation is facilitated by the use of the Touchtone keyboard. AT & T added two extra keys to their Touchtone telephone, as shown in Fig. 11.1. Now they make only 12-key, not 10-key, telephones. The added keys labeled ✳ and # are for the purpose of facilitating data transmission and other functions such as Picturephone dialing.

Much of the United States can use Touchtone telephones today; it is intended that almost all of it will be able to by 1975. For areas that do not yet have Touchtone dialing, a small, extra non-AT & T keyboard can be added to a dial telephone set as in Fig. 11.2. Eventually dial telephones may cease to exist in the United States and all equipment would be Touchtone.

In this chapter we shall discuss to what extent the Touchtone telephone can be used as a computer terminal with no additional equipment except at the computer center. As we shall see, all types of interesting and useful applications could be constructed around it. The offering of such services will enable the public to "dial" a computer from their own home and use it at very little cost. If this use of computers becomes accepted in the home, the salesman of more expensive terminals will already have his foot in the door.

135

Figure 11.1. The Touchtone telephone keyboard has twelve keys rather than ten. The two nonnumeric ones, "✱" and "#", facilitate its use in many data transmission applications.

Figure 11.2. Where Touchtone telephones are not yet in service, a small separate Touchtone keyboard can be added to the telephone, as shown.

VOICE ANSWERBACK   If the telephone alone is to be used as a computer terminal, the computer must respond with human-sounding voice answerback. This step is satisfactory in many applications, although, for some, a written or displayed response would be much better.

A variety of voice answerback devices exist on the market. Several are inexpensive and have a vocabulary of a small number of fixed words. Others are parts of computer systems with potentially very large vocabularies.

In general, there are three ways in which the human voice can be stored in machines.

1. In *an analog form*. It can be stored as on a phonograph disk, magnetic drum, on recording tape or photographic film, perhaps in fixed-length words. A simple electronic circuit can transfer it to the telephone line [1].

2. *In a directly digitized form*. As discussed in Chapters 6 and 17, one second of sound is represented by 56,000 bits in PCM telephone transmission. These bits could be stored on computer disks or other storage. A computer program could select and connect together different variable-length strings.

3. *In encoded digital form*. Instead of storing actual sound "samples," as with PCM transmission, the words can be encoded. A sequence of digits represents the timing, energy, and frequency associated with the spoken words, in a compact form. Special circuits must then reconstruct the sound of the word in order to speak it. Spoken words can be encoded in various ways. Typical schemes would require less than one-twentieth of the bits that seven-level PCM encoding requires. They would give clearly intelligible words but not reproduce inflections exactly or necessarily permit the speaker's voice to be recognized, as would PCM. However, the whole human vocabulary could be stored on a computer disk storage unit. The speech heard with this method can be less staccato and zombielike than method 1, for the words are stored and strung together in a variable-length rather than a fixed-length form.

An inexpensive example of the first method is the Cognitronics "Speechmaker," which uses a 31-word vocabulary in the form of a photographic film drum. A five-bit code is used to select one of the 31 words, or silence. The drum is easily interchangeable. Several drums with prerecorded vocabularies are available, and drums with new words can easily be made. The sound is generated by a light beam shining through the film onto a photoelectric cell. The output of the cell is amplified and put onto a telephone line.

The drum rotates every 625 milliseconds, and each word (or segment of speech) lasts 600 milliseconds. The drum generates a timing pulse once per revolution, and in the next 25 milliseconds the switching to the selected sound track takes place. The circuit remains switched for the 600 milliseconds of playing time.

For many applications, it is desirable to have longer responses—for example, complete sentences like "Please key in your account number" or "Press the asterisk key and repeat your last message." It would seem worthwhile to have two voice drums, one with a 600-millisecond time slot and one with a 6-second time slot.

An example of the third approach is the IBM 7772 [2]. With this machine, the words are stored on disks or other computer storage. Table 11.1 shows the lengths of typical words and the number of eight-bit

Table 11.1 THE STORAGE REQUIREMENT FOR TYPICAL SPOKEN WORDS IN THE IBM 7772 VOICE ANSWERBACK SYSTEM*

Spoken Word	Length (in milliseconds)	Number of 8-Bit Bytes Used for Storage
One	466	139
Two	447	137
Three	472	145
Four	554	122
Five	686	154
Six	567	148
Seven	524	129
Eight	456	157
Ten	461	148
Accrue	676	229
Accrued	772	262
Action	536	223
Actual	600	233
Affiliation	923	259
Allot	581	207
Allowable	779	210
Allowed	642	186
Analysis	776	233
Appropriate	746	303
Appropriation	950	391
Asked	601	190
Assistance	703	259
Assumed	703	211

*Note that it is much less than the storage needed for PCM encoding of the same word. Information taken from reference 3.

bytes of storage they require [3]. One second of sound requires about 2400 bits of storage. The computer, in answering an inquiry, feeds the bytes representing the spoken response to the 7772. A voice-code translator converts them into sound and switches them to the requisite telephone line. Figure 11.3 shows a 7772 configuration. The words could

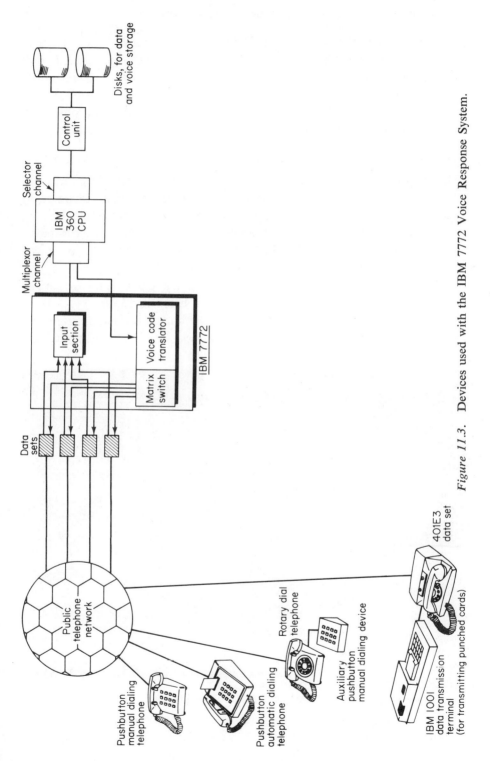

Figure 11.3. Devices used with the IBM 7772 Voice Response System.

139

be in any language or accent. The IBM sales manual used to say "Female voice, no extra charge." Now the vocabulary normally provided is that of a female voice!

The Touchtone telephone can be used as a terminal with any one of these machines. As indicated in Fig. 11.3, a variety of additional keyboards and card readers can be added to the telephone if required.

PLASTIC OVERLAYS   In order to use the Touchtone telephone as a terminal, its keys must be specially labeled. Each key must be given a second meaning in addition to the digit on its face. If this use of the telephone catches on, it is probable that many different labels will be needed in order for different computing services to be dialed. The labels would probably be in the form of an interchangeable plastic overlay. The user may have a booklet of such overlays, each one having written on it the telephone number of the computer that should be dialed when using it. Figure 11.4 shows such an overlay. The normal Touchtone telephone has a flat area around the keys on which such an overlay can be positioned. The overlays used

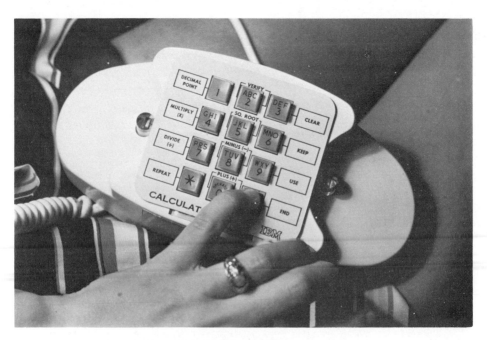

Figure 11.4. Touchtone telephone with overlay.

by the author for demonstration purposes were made with a Thermofax copier using "Color Accent" transparency film. They proved very satisfactory.

Let us now discuss applications of Touchtone telephones using such overlays.

DESK   In 1966 IBM experimented with a New York
CALCULATOR   City high school, giving the children a keyboard like a Touchtone telephone to add to their existing dial telephones. This keyboard permitted them to use the distant computer as a remote calculator. They could use it at home for their homework. The scheme was an instant success. The pupils quickly learned how to use the device and enjoyed playing with it. The system could do any calculation that can be done on a desk calculator and could work out square roots with ease. If the user made a mistake at the keyboard that the computer could detect, it would inform the child, again with a clear human voice. An IBM 1620 was used, an inexpensive second-generation computer.

The conversation used with this scheme is described in *Design of Real-Time Computer Systems*, Chapter 8. A more versatile scheme has been devised at Houston University, Texas [4]. This scheme permits the use of any programmable functions and allows the user to enter loops and call subroutines as he would do in a computer program. It is somewhat more difficult to use than the IBM schoolgirl scheme. Perhaps when such schemes are eventually marketed, it will be worthwhile to have both a simple desk calculator and an elaborate calculator that would appeal to a person familiar with programming.

With the Houston scheme, the Touchtone telephone may be labeled as in Fig. 11.5. Each operation entered with the keyboard ends with the pressing of the $\#$ key. Thus numbers may be added, subtracted, multiplied, and divided by pressing key as follows:

$+$ 3.79 $\#$
$-$ 27.34 $\#$
$\times$ .4 $\#$
$\div$ 3.7865 $\#$

The computer can be instructed to speak the answer obtained by pressing the key labled "*Answer*," followed by $\#$. If so desired, intermediate results can be read out

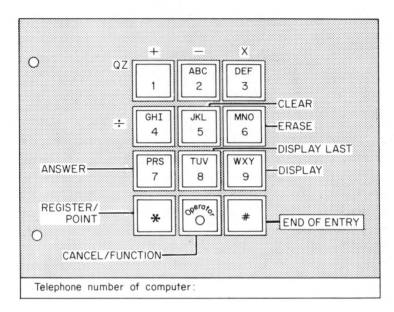

Figure 11.5. Overlay for labeling Touchtone telephone for desk calculator operation [4].

as the calculation proceeds. *"Display"* causes the computer to read out the instructions it has been given.

Quite complex programming can be carried out with the telephone in this system. *Subroutines* can be constructed and *functions* stored. *Registers* can be designated and used. For example, suppose that it were required to execute the following PL/I DO loop:

```
A = O ;
DO N = O TO M BY Y ;
A = A + Y * * 2 ;
END ;
PUT DATA (A) SKIP ;
```

The user would proceed as follows where M is in Register 1 and Y is in Register 2:

```
Clear 4 #
Function LOOP # 9 # 0 # Register 2 # Register 1 #
Clear 3 #
+ Register 2 #
X Register 2 #
+ Register 4 #
Clear 4 #
```

```
+ Register 3 #
Function END #
Answer Register 4 #
```

A is here accumulated in Register 4, and the last *Answer* statement causes the result to be spoken.

Most applications require a far less complex use of the Touchtone keyboard than this, but it is interesting to note that complex programming can be done without too much difficulty.

VARIETIES OF APPLICATIONS Voice answerback has a great many potential applications both for commercial and domestic users. One manufacturer's sales brochure lists the following commercial uses [5]:

Manufacturing Industries
> Order entry
> Job status
> Parts inquiries
> Inventory control

Retail Stores
> Credit inquiries
> Stock status
> Merchandise control
> Order inquiries
> Credit and collection inquiries

Hotels and Motels
> Reservation inquiries

Hospitals
> Laboratory report inquiries
> X-Ray results
> Patients' characteristics
> Patients' locations

Insurance Companies
> Policy inquiries

Educational Institutions

Substitute teacher assignments

Banks and Financial Institutions

Deposit accounting inquiries
Installment and commercial loans
Savings account inquiries
Brokerage inquiries

In addition to private commercial applications, a variety of uses may appeal to the domestic user. For example, he could find out the latest stock prices and ask for other stock information that may interest him. His home telephone thus becomes equivalent to the terminal a stock broker uses. Similarly, he could request movie and theater information, perhaps making bookings from the home. He could ask for plane, train, and bus times. New and appealing transportation schemes have been proposed involving computer-scheduled buses that pick up customers from their homes or offices with completely flexible routing. After dialing for the bus, the customer could communicate with its scheduling computer via a Touchtone telephone [6].

Diet planning and menu selection have also been proposed. Job seekers could use the system to search for new employment. A voice answerback system could be used to obtain hotel rooms, domestic help, library books, magazine subscriptions—in fact, almost anything from attempting to buy a second-hand car to real-time computer dating.

The applications can become quite complex, as in the case of elaborate uses of the Houston desk calculator program. North American Aviation used such a scheme to allow engineers to inquire about engineering drawings. A file of 75,000 or so drawings was searched and the latest changes listed. The Equitable Life Assurance Society used voice answerback in the processing of insurance applications.

In Detroit a real estate system was developed by the Realatron Corporation. Several thousand participating brokers enter information about the location, price range, style, number of bedrooms, and up to fifteen other preferences relating to the houses they are interested in, and the computer then searches its files. In addition, a variety of credit-checking and banking operations have been done with Touchtone keyboards.

In New York and Illinois, Touchtone telephones have been used for teaching. The New York City Board of Education has used a system that can drill up to 2000 students, in grades 2 through 6, in arithmetic. The student uses the Touchtone telephone at home. The computer poses voice questions to the student, who answers at his own rate; the computer then notes the responses. In Oak Park-River Forest High School near Chicago a computer instructs and drills telephone students on subjects as varied as history, biology, languages, business education, mathematics, and the physical sciences. The instruction is from audio lessons taped on continuous loops of 1-in. tape. Each loop contains 32 different sound tracks up to 15 minutes in length.

Where a number of plastic overlays are used to enable a user to carry out different applications at his terminal, it would be useful to standardize some of the operations. Figure 11.6 shows a blank overlay; Figs. 11.7 to 11.10 show overlays marked for several applications that might apply to a home user. The user might have a booklet of such overlays. Each overlay contains the telephone number of the computer that must be dialed with it. An "application number" tells the computer which overlay is on the telephone. The number is sent to the computer by pressing the 0 key (labeled "application number: XXX") and then keying the number.

The * and # keys are labeled "REPEAT" and "END OF ENTRY." The latter is pressed at the end of each item keyed in. The former is

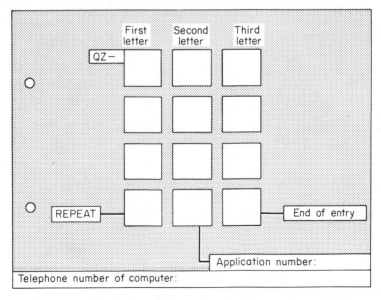

Figure 11.6. Blank overlay for labeling Touchtone telephone for diverse applications. (See Figs. 11.7 to 11.10.)

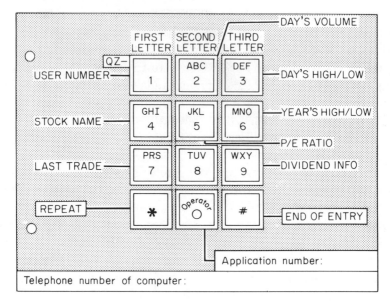

Figure 11.7. Overlay for labeling Touchtone telephone for stock market inquiries.

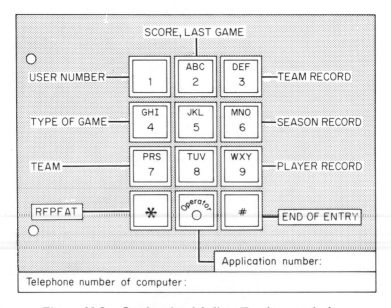

Figure 11.8. Overlay for labeling Touchtone telephone for obtaining sports information.

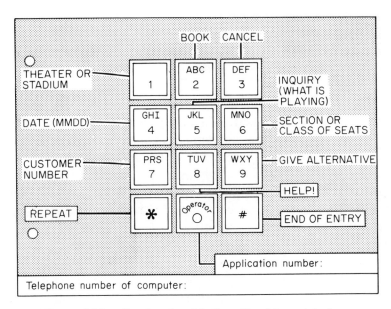

Figure 11.9. Overlay for labeling Touchtone telephone for theater/stadium inquiry/booking service.

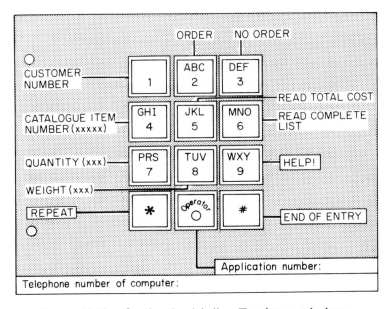

Figure 11.10. Overlay for labeling Touchtone telephone for home catalog ordering service.

pressed when the computer has spoken a response and the user wants it repeated because he may have missed something.

Figure 11.7 shows a template labeled for obtaining stock market information. Figure 11.8 shows one for sports information. In the former, the user must know the three alphabetic characters that are used to reference each stock. In the second, he must have a listing of numbers referring to teams and players. The question arises, How will a user be billed for these services? He could be automatically billed by the telephone company every time he dials the number of the computer in question. Alternatively, he might pay once a quarter, perhaps, for the privilege of using the service and would then be given a self-checking user number that could be registered in the computer. Information could then only be obtained after keying in the user number. A better method would be to make each telephone set automatically generate a unique signal number after the key labeled *User Number* has been pressed, which is done on some terminals. In the future, computer voice recognition might become effective enough to use for verification. The best way to solve the billing problem would be to have it handled automatically by the telephone company.

The next two figures show templates in which the user is requesting a service. Here the billing problem does not arise because he is billed only for the service he receives, which normally involves something mailed or delivered to his address. Mere inquiries are free, although again a customer number is required. Figure 11.9 shows a template for inquiring what is playing at the theaters, concert halls, and sports stadiums, and for making bookings. Such a service might be provided by a theater agent. The user will key in the number of seats he wants and the section or class. The machine will read out appropriate seat numbers, and he will inform it if he wishes to book them. He can request alternatives and can cancel seats if he wants. The tickets will be mailed to the customer if there is time; otherwise he will pick them up at the office or theater.

Figure 11.10 shows a similar scheme for ordering goods. Such a service might be provided by a merchandizing organization. For each item number keyed in, the computer will read back what it is as a check and will ask the user what quantity or weight he requires. The computer will then quote a price. The user will press the *Order* or *No order* key, depending on whether he wants it. At any point he can ask for the total price so far or for a read-out of the complete list he has ordered.

Any such schemes would probably use self-checking account numbers to lessen the likelihood of bogus access to the system. For security reasons, it would be advantageous to have "voice print" checking by computer to confirm the identity of the caller, and this is now being experimented with.

Both Figs. 11.9 and 11.10 have a key labeled Help. Pressing this key gives the caller access to an operator who can help in any difficulties encountered. The # key either terminates an operation or terminates an entry, as in the Houston desk calculator case.

It is necessary to transmit alphabetic characters in many applications. Three alphabetic characters are normally used in obtaining stock quotations, for example. On a Touchtone keyboard these must be entered in a two-character form. An easy way to do so is to first press the key labeled with the letter in question and then press 1, 2, or 3, depending on whether the user wishes to transmit the first, second, or third letter on that key. Thus pressing 5, then 2, would mean K. Unfortunately, Q and Z are missing from the Touchtone keyboard. These letters and - have been added to key 1 for this illustration and are marked on the overlay.

OTHER DEVICES USING TOUCHTONE FREQUENCIES   The Touchtone telephone uses eight frequencies which are now a standard of the United States telephone networks (see Fig. 11.11).

The frequencies are arranged in two groups of four. Group A relates to the rows of keys and group B relates to the columns. The pressing of any key on the telephone results in its two frequencies—one from each group—being transmitted. This process would permit 16 keys to be used on a telephone, and indeed a fourth column of keys is found on military telephones. These extra keys have not been used as yet on civilian telephones.

Using the eight frequencies, coding schemes can be employed that would permit more than 16 keys. Using binary encoding, 256 different characters could be transmitted. The scheme using two frequencies per key gives some measure of protection against stray signals on the line and voice-generated tones. If one frequency is picked up without an appropriate associated frequency, it is ignored. Parity checking could similarly be used. If there must always be an *even* number of two or more frequencies per character, this scheme would permit 127 characters. A 4-out-of-8 code, in which each character must have exactly four frequencies, would be somewhat more secure than parity checking and would give 70 possible characters.

Alphanumeric or other keyboards can be built using such coding schemes. Figure 11.12 shows a portable IBM terminal with 60 keys and

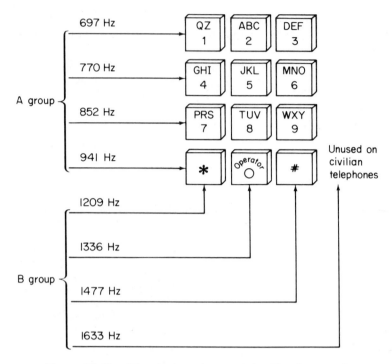

Figure 11.11. The frequencies used by Touchtone telephones. Each key transmits one frequency from each group.

a loudspeaker. It has an acoustical coupler and thus can be used anywhere, including public telephone boxes. The terminal transmits a unique identification number that enables the computer to identify which terminal it is. The circuit for transmitting this number is reasonably tamperproof; hence privacy and security can be ensured in an application. Plastic overlays can be used to label the keyboard.

REFERENCES

1. *IBM 7770 Audio Response Unit, Component description manual,* No. A27-2712, IBM Corp., Kingston, N.Y., 1969.
2. *IBM 7772 Audio Response Unit, Component description manual,* No. A27-2711, IBM Corp., Kingston, N.Y., 1969.
3. *Vocabulary File Utility Program for the IBM 7772 Audio Response Unit.* Manual No. C27.6924–2 IBM Corp., Kingston, N.Y., 1969.

4. Albert Newhouse and Robert A. Sibley, Jr., Houston University, Clearing house for Federal Scientific and Technical Information, U.S. Department of Commerce, National Bureau of Standards, Document No. AD 691398, 1969. "On the Use of Very Low Cost Terminals."

5. *"For Spoken Replies from Your Computer,"* IBM brochure No. 520–1965–2, IMB Corp., White Plains, N.Y., 1965.

6. James Martin and Adrian Norman, *The Computerized Society* (Englewood Cliffs, N.J.: Prentice-Hall, 1970), Chap. 10.

Figure 11.12. The IBM 2721 in use in a public call box. The 2721 is an inexpensive, light, portable voice answerback terminal with 60 keys and the ability to transmit a code number for security purposes.

1922. A picture from the Illustrated London News showing a family listening to an early broadcast. One of the pioneers of radio—a man who later became very famous—had said that he could foresee little use for public broadcasting except for the broadcasting of Sunday morning sermons because that was the only form of mass oratory that the public regularly listened to.

12   TERMINALS IN
        THE HOME

With the exception of color television, the last decade has brought surprisingly few new types of inventions for the home. This deficiency will probably be made up in the next decade or so; all kinds of devices based on computer logic circuits, telecommunications, memory devices, and videotape are presently envisaged. An executive of Texas Instruments [1] estimates that homes in the United States may soon contain as much as $10,000 worth of electronic devices. An article in *Forbes* [2] predicted a $30-billion a year market for such equipment, thereby challenging the auto industry for the largest single share of consumers' income by 1980.

Besides improvements in passive entertainment media, it seems likely that terminals which interact in some way with a computer will be introduced into the home. Acceptance may be slow at first, but eventually this innovation will become a mass market.

Perhaps one of the factors leading to the introduction of terminals into the home will be the capability to communicate with a computer by using an ordinary key telephone as described earlier. As an adjunct, to this it may become possible to add a slow and inexpensive printer to the telephone connection. The current Sears Roebuck catalogue has an electronic calculating machine with a printer for only $58.95. With a little more electronics this could become a crude terminal. Another likely possibility is the use of an adapter that causes responses from the computer to be displayed on an ordinary television set, as in Fig. 12.1. One such adapter has been built by Computer Transceiver Systems, Inc., New Jersey.

For persons who do not have Touchtone telephone service, a separate keyboard can be purchased. One such keyboard, which is made by the Metroprocessing Corporation of America at a cost of about $200, can be attached to any dial telephone through a coupler that fits onto its mouthpiece. The user could sit in his favorite chair facing the television, with the keyboard in his lap. It weighs 2½ pounds and measures $2 \times 4 \times 8$ inches. More elaborate home terminals will undoubtedly follow these devices, especially for commercial use.

What would be the purpose of having terminals in the home? There are many possibilities—business, entertainment, education, household services. Their main use today is for business rather than entertainment or personal services.

The introduction of such terminals will probably be phased—starting with industrial users and then going to amateur enthusiasts, thereby paving the way for the general public.

HOME TERMINALS USED BY INDUSTRY   Computer terminals in homes *today* are generally not there at the user's expense. These users, for the most part, are university staffs engaged in research, programmers employed by a large corporation or sometimes a "software house," a handful of executives enthralled by the new technology, and some traveling salesmen who use their terminals when they arrive home to transmit their orders to a computer.

The first widespread use of home terminals will probably be sponsored by employers who expect to benefit thereby. Certainly the author could make good use of a home terminal installed by his company, and there are thousands of systems engineers, programmers, and technical staff who could similarly benefit. What would they use them for? Writing programs, developing their own collection of programs for systems analysis, design, or other work, obtaining technical documents, looking up facts, running teaching programs, and exercising newly learned skills. They, and their children, may decide they need to learn a new computer language. For several evenings they may spend an hour with a manual and at a terminal using a teaching program. However, the only way *really* to learn a computer language is to program in it. So they may dial up a computer that enables them to program with a given language "on line" and then spend a period each evening developing their ability to use the new tool.

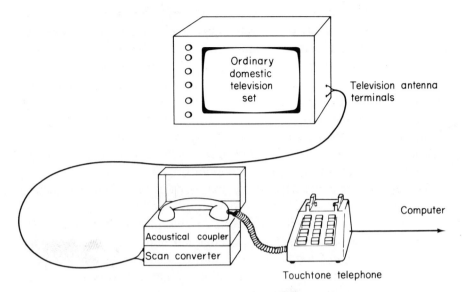

Figure 12.1, Digital signals from a computer, transmitted over the telephone line, can be displayed on the screen of a conventional television set by means of an inexpensive acoustical coupler and scan converter. The user responds to the computer with the keys of the Touchtone telephone.

Many technically minded people would find it pleasant to sit down for an hour or so after dinner and tinker with a program they are working on. Often, too, ideas come at strange times—when the mind is relaxed. One wakes up in the morning and suddenly sees how to do something that was puzzling one the night before, or realizes where the cause of an error lies. So one would wander up to the terminal, in one's pajamas, and check the thought or modify the program there and then.

In writing a report, a group of authors could type the text directly into their respective home terminals. The report would reside in the memory of a distant machine. They could then modify it, edit it, restructure it, snip bits out, correct each other's work, add to each other's ideas, and instruct the terminal to type clean copies when they were ready. A team of authors in Washington is writing a technical report in just this way. Possibly magazine editing will be done with such aids in the future.

When the opportunity to use computers from the home becomes common, it seems logical for some employees to spend at least part of their time working at home rather than at an office. The overhead cost of providing staff with offices and desks is very high, especially in large cities; some of this cost will therefore be saved by the use of home terminals. A manager can see what remote staff members are doing by telephoning them, dialing up the computer they are using and examining their work, and sending them instructions on their home terminals. It is possible, indeed, that in the future some companies may have almost no offices. A software company for producing programs may cut its costs significantly if most of its personnel work at home. Mothers who participate in such a scheme may be relieved of much of the boredom they feel when they are unable to leave their children.

The growth of computerized teaching—from today's experiments to tomorrow's industry—will need a tremendous amount of human thought and development of programs. Much of this endeavor may also take place in the home. Step by step, a teacher can build up his lessons on his home terminal. Occasionally he may dial a colleague to ask him to try out what he had produced on *his* terminal. When the work is near completion, he may try it out in a classroom, study the reactions of students, and return home to make appropriate modifications.

This is not only possible for teaching programs, it is also possible for many other computer uses now within our grasp. Legal data banks, medical data banks, data banks for all types of professional users, as well as nonprofessional people seeking information, are going to take an enormous amount of intelligent effort to build up. Much of this work is likely to be done with terminals, probably terminals which impose standardized formats on the data being entered. It is as though we have to rewrite all our textbooks and reference documents in a form that permits terminal users to access the data and instruct computers to search for and manipulate the information.

We see here a beginning of a return to "cottage industry"—a trend that will probably increase greatly during the next few decades. Nevertheless, at the moment, a few factors are counteracting this trend. The first is the reluctance of some companies to give their systems analysts or other employees home terminals because such a step seems an unprecedented and potentially unpopular encroachment on leisure time. The second is a feeling that regardless of what they are doing, employees ought to be at their desks from 9 A.M. to 5 P.M. (at home, who knows, they might be watching television!). The third is a feeling that people cannot work at home because the environment is not suitable. Perhaps some homes are unsuitable, but the author of this book had no difficulty in writing it at home with a rate of productivity that would be regarded in industry as high. We believe that these three views are not generally held by intelligent and enthusiastic members of the community and that this is an era when we must throw off the mores of tradition and rethink what is applicable to the age of teleprocessing.

Certainly if such "cottage industry" spreads extensively in our society, it will drastically change social patterns.

COMPUTER AMATEURS   Perhaps the next group to help introduce terminals into the home on a large scale will be the computer hobbyists. The equipment that ham-radio enthusiasts and other such groups now have at home is more expensive than basic computer terminals, such as teleprinters or alphabetic screen units. Furthermore, the interest and excitement likely to arise by being able to dial up and work with an ever-growing number of computers will probably be far greater and far more absorbing than the attraction of such hobbies as ham radio and home movies. It is probable that as the number of available computers grows, as education on computers spreads, and as leisure time increases as a result of automation, the computer amateurs will become a growing body. Magazines will be produced for them. Industry will encourage them and enthusiastically sell to them.

Computer hobbyists may fall into a number of different groups. Some will hope to produce and sell their own programs or make them available to other amateurs. There will be some who are less creative and mainly interested in education; they will dial various instructional and information-retrieval programs. Others will be interested in playing games, doing puzzles, or indulging in mathematical recreations. But perhaps the majority will fall under the narcotic spell of programming. Working on ingenious programs (rather than the routine of commercial programming) is, to a certain type of mind, endlessly captivating.

The computer amateur will have significant contributions to make to the development of this technology. Most technical fields are too complicated or specialized for the amateur to make a name for himself, but programming new computer applications has endless scope. In every direction, new territories await the ingenuity and care of a dedicated amateur.

As noted earlier, *enormous* quantities of ingenuity and programming are essential to this new era—not the work of geniuses but ordinary step-by-step construction and testing. The work requires a high order of craftsmanship. In general, it is creative, enjoyable work, work that wives and children can do, work that disabled and in some cases blind people are doing. It is work to which the hobbyist or the enthusiast making money in his spare time will contribute enormously.

DOMESTIC   The time will come when the computer termi-
MASS MARKET   nal is a natural adjunct to daily living. Sooner
                  or later computing will become a mass domestic market and the computer manufacturers' revenue will soar. The airline industry, the automobile industry, telecommunications, and other complex technical industries all spent two decades or more of limited growth but then expanded rapidly when the general public accepted and used their product. Data transmission is going to make this process happen in the computer industry also.

The eventual uses for terminals in the home are endless, and numerous computer systems will be set up for domestic and entertainment purposes. Whether this will come about in the next 10 or 15 years is difficult to tell. We have the technology today, although innovations ahead will drastically drop its cost. Let us consider the possibilities, anyway—after all, the computer industry moves with surprising speed.

The home user will in time have access to a wide variety of data banks and programs in different machines. He will be able to store financial details for his tax return, learn French, scan the local lending library files, or play games with a computer. If he plays chess with it, he will be able to adjust its level of skill to his own. It has been suggested that news will be presented via terminals in the future; the user will skip quickly through pages or indexes for what he wants to read on his screen. Because the machines' files will be very large, foreign newspapers transmitted by satellite could also reside in his local machine. He may have a machine to *print* newspapers in the home, although use of the screen might be preferable. If he wants a back number, he will be able to call for it, using a computerized index to past information.

As illustrated previously, a man interested in the stock market could dial up a computer holding a file of all stock prices, trading volumes, and relevant ratios for the last 20 years [3]. Possibly stockbrokers would make the information available free and would provide analytical routines to their clients. When a client bought or sold stock, he would be able to give the appropriate orders directly to the stockbroker's computer via his own terminal.

On the other hand, someone lacking the money to buy stock could *play* at buying it. The computer would calculate the effect of his buy and sell orders, permit him to trade on margin, pretend to give him loans, but no actual cash transfer would take place (apart from the cost of using the machine). When friends visit, he would be able to dial up his records to show them that he started with $100,000 and in six months had made $78,429! Perchance to dream in glorious detail. At least he will have had lots of practice ready for when he does eventually become rich!

SHOPPING   Although a man might use a home computer terminal for scanning newspapers or for stock market studies, his wife is more apt to use it for shopping. In some countries, punched-card supermarkets have come into operation. The housewife at the supermarket picks up a card for each item she wants to buy and these cards are then fed through a tabulator. She pays the bill, and the goods are delivered from the stockroom. The advantages for the supermarket are that less space and less capital outlay are needed and there is less pilferage. Still, why must the housewife come into the store at all? She could scan a list of the available goods and their prices at several different shops on the home terminal and then use the terminal to place the order. An organization selling in this way could cut overhead to a minimum by eliminating stores and lessening their size.

The customer who buys through such an automated catalog avoids the exasperation of fruitless searches for hard-to-find merchandise. Besides replacement bulbs for projectors condemned by planned obsolescence, spare parts for automobiles, or rare phonograph records—all items likely to be found, if at all, in only one store in the city—there are tedious searches for special items like summer houses for rent, theater tickets for a particular show, or boats with a particular specification. Presently these items are found through classified ads in newspapers or through agents.

Much of the work of such agents, and of the advertisement columns, could be done more cheaply and efficiently by a data bank accessible from terminals. Until home terminals are common, it is unlikely that those who now make their livings from putting buyers in touch with sellers will cooperate with each other to set up systems that will make themselves redundant. A British service to homeseekers foundered when most real estate agents refused to place details of properties on their books in the computer system as well, even though the purchaser would have had to use the agent to reach the seller. Theater ticket systems are doing better.

Many newspapers and magazines, already competing with television for advertising revenue, may go out of business. Printing unions are well organized to prevent the automation of printing itself; thus many newspaper publishers may be unable to avoid closing down when the classified ads emigrate to a computer.

The preceding services may be provided free by the advertisers; many others will have a charge, with the bookkeeping almost certainly being done on a central computer. An automated diary might cost one cent per entry—a small fee to ensure that no appointments are missed and no birthdays forgotten. Hunting through dictionaries, almanacs, abstracts of literature, and similar sources might cost about one dollar an hour, much less than a trip to a good library, and more fruitful.

As more such services become available, the economic justification for terminals in the home will be strengthened. Nieman Marcus, the world-famous department store in Dallas, Texas, catering to people to whom economics are irrelevant, featured the Honeywell H316 in its 1969 Christmas catalog, ready programmed with menus, diets, and routines to handle domestic accounts. The cost was outrageous, but in the future terminals will add such services to their other functions at minimal cost.

SPORTS   Sports, too, will be aided by computer, and the sports fan will be able to obtain all manner of information on his home terminal. In top golf tournaments today, computers are used to keep track of everything that is happening. Observers stationed around the course report information to a computer station by means of walkie-talkies. The machine digests all the information, operates a scoreboard for the clubhouse gallery, press, and television; and displays on a screen hole-by-hole scores and such information as greens reached in par, number of putts on each hole, and lengths of drives on selected holes. Instantaneous comparisons between players can be produced along with all manner of asides, such as remarkable runs of birdies. And golf, as one of my reviewers insisted that I comment, does not come up to baseball in providing useless comparisons.

The terminal owner of the future will presumably be able to dial machines giving up-to-the-minute detailed information on any sport instead of being restricted to the one or two items fed by the television channels. One imagines a fan of the future watching pro football television on a Sunday afternoon much as today, but with the teleprinter chattering away at one's side, printing the commentaries it has been instructed to give on other games.

HORSE
RACING
In the winter of 1967–68 England was swept by a disastrous plague of foot-and-mouth disease that killed many millions of pounds-sterling worth of livestock. Most horse racing was stopped because of the epidemic, and workingmen who enjoy placing bets with local bookies were highly despondent. Fortunately, *The Evening Standard* of London realized that betting on a horse race does not necessitate real horses galloping around a track—indeed, in the age of computers it is inefficient and a waste of manpower (to say nothing of "horsepower"). Consequently, the paper made the following announcement [4]: *"The Evening Standard* today proudly announces that it has devised, and will stage, the World's First Electronic Horse Race." It went on to say that the Massey-Ferguson Gold Cup, canceled because of foot-and-mouth disease, would be run on the London University Atlas computer. A mathematical model of the horse race was programmed with the help of racing experts who provided the details on horses and their form over the previous two years, jockeys, distance between fences, and so on.

The computerized horse race met with the full approval of the National Hunt Committee and the Cheltenham Racecourse Executive. The BBC broadcast a full commentary on the race, with commentators sounding no less excited because the horses were not real. The commemorative Gold Cup was awarded to the winner; the jockeys received their normal fee.

Clearly this concept can be extended. With terminals in the home, the racing enthusiast can have a race any time he wishes. It would probably be a fine after-dinner entertainment. He and his guests could use the terminal to ask questions about the various horses' form, and the computer might ask whether these are simulated bets or whether an actual cash transfer will take place. The race should be no less exciting than an actual race. There is no need to have the monotonous voice we sometimes hear on these machines. Even today's equipment can reproduce the voices and even the intonations of popular sports commentators. Using cable TV in an interactive fashion, as discussed in Chapter 5, bets could be placed as the race takes place, and in other simulated activities the viewer might be permitted to modify the course of events.

Many other such games will be played with the terminals. Who knows what forms gambling might take in the computerized society, with the home-gambler's bank balance being automatically added to or depleted.

Perhaps in America one will be able to telephone the local Mafia computer.

HOUSEHOLD
APPLIANCES

The telephone line, in addition to its normal use, could be used for activating household appliances. A family driving home after a few days vacation, may be able to telephone their home and then key some digits on the Touchtone telephone that switch on the heat or air-conditioning unit. Before leaving for work a woman will preprogram her kitchen equipment to cook a meal. She will then phone at the appropriate time and have the meal prepared. Or possibly a computer might telephone the equipment and instruct it step by step to perform a sequence given to the computer the night before—for example, to switch on the oven at 3 P.M. to cook the roast that had been placed in it and to move some vegetables out of the freezer compartment, leaving them to thaw. At the appropriate time, the vegetables would be heated in their aluminum foil, and the dish-warmer would be switched on.

TELEPHONE
LINES

What communication links would be used for home terminals? When a computer terminal is used in the home today, it is almost always connected by means of a conventional telephone line. Two wires are permanently connected from the home of each subscriber to the local central office; with the exception of party lines, these wires are not shared. It is staggering to reflect that in Manhattan, for example, there are two wires under the street for almost every private subscriber. Ninety-nine percent of the time the author's telephone wires are idle, as are most people's. Furthermore, when in use these two wires carry less than one-twentieth of the voice traffic they are capable of carrying. It is rare indeed in engineering to find such an expensive facility used so inefficiently. When coding in the form of data is considered, they carry only a minute fraction of what is possible.

Without any change in telephone lines, therefore, we could use terminals in the home if the appropriate computer's telephone lines were attached to the same local exchange (central office). An expansion in the central-office equipment would probably be needed. The lines would be in use for a longer period of time, and thus the number of simultaneous paths through the exchange would have to be increased somewhat. In general, however, there is no *technical* reason why our telephone bill for local computer calls should be much higher than it is today.

Long-distance calls to a computer are a different matter, for they do not involve a permanently connected pair of wires; thus we can expect an increase in cost proportional to our increase in line usage. Some terminal uses in the home will occupy the line for lengthy periods. Consequently, such densely populated areas as New York or London the preceding schemes will not involve a high line cost because of the proximity of the relevant computer. For areas far from the population centers, line cost will be high. It appears, then, that initially there will be a marked line-cost advantage to living in a metropolis. However, computer services will slowly spread to smaller towns, and as mentioned in Chapter 3, the cost of long-distance channels is dropping fast. A long-distance voice channel may drop to one–tenth of present cost in ten years; and with efficient data organization, many terminal users may be routed to a computer over one voice line [5].

OTHER TYPES OF CHANNEL   Several other ways of connecting home terminals exist. With today's tariffs, private-leased lines—voice grade or subvoice grade—may be the cheapest method if the terminal has a high utilization. This fact is especially true for lengthy lines with more than one terminal attached to one line by means of multiplexers or concentrators [5]. Computer programs exist for designing minimum-cost networks of this type [6].

Leased lines may, however, restrict the versatility of the user. He may want to dial a wide variety of different computers, which he could do on the telephone network. In the years ahead it may become cheaper for him to subscribe to a switched data service, such as those discussed in Chapters 9 and 10. The Datran System, for example, would give him cheaper rates than the public telephone network for calls beyond his local area.

As discussed in Chapter 5, the coaxial television cable offers another possibility for connecting terminals. Such a cable may have a very high speed data channel traveling in a closed loop. One data channel may carry information both to and from the terminal, or there may be separate channels for data to and from the terminal. If there were a sufficiently large number of terminal users in a small geographical area, a cable carrying a digital bit stream would be cheaper than the telephone network for connecting them.

For the domestic user, the television cable gives the added attraction that *pictures* can be received, in color, as responses in conversations with computers. This process will have many appealing applications. It will be particularly valuable with teaching programs for children. Children usually find computer terminals a fascinating toy and want to play with them endlessly. They will have no difficulty in learning to operate the keyboard that enables them to converse with this new medium.

In the advertisers' paradise of America, all kinds of highly colored catalogs will become available the moment the new medium arrives, and there will be varied enticements for exploring them. Very elaborate presentations of products will become possible. The Sears-Roebuck catalog, for example, might include film sequences, although the user would still be free to "turn the pages," to use the index, to select and reject. As with American television, advertising would help to pay for the new medium. Perhaps critical consumer guides will also become automated to aid product exploration. Having scanned the relevant catalogs and inspected pictures of the goods in detail, the shopper could then use the same terminal to order items, with the money being automatically deducted from his bank account.

Picturephone will also offer the possibility of a video channel into the home. As with telephone lines, a subscriber will have his own private Picturephone cable, which nobody else uses. Most of the day it will lie idle. The cost of a *local* Picturephone call to a computer at the subscriber's central office may therefore not be high once the wiring exists. On the other hand, lengthy long-distance calls using the full Picturephone bandwidth will be very expensive. A metropolitan computer may offer interesting local services via Picturephone. The low resolution of today's Picturephone screen, however, is a disadvantage. Experiments using video-phones for shopping have been unsuccessful because of the low resolution [7]. For many domestic applications, a higher resolution at the price of less capability to follow movement would be preferable. Facilities of lower cost than Picturephone are needed for the initial acceptance of terminals in the home.

In parts of this chapter, we have already forecast a future that may be beyond the next decade. However, before we look further at this more distant future we should discuss some of the technology in more detail. This is the function of the next section.

REFERENCES

1. *Forbes.*

2. *Forbes,* March 15, 1967.

3. James Martin and Adrian Norman, *The Computerized Society* (Englewood Cliffs, N.J.: Prentice-Hall, 1970).

4. *The Evening Standard*, London, December 9, 1967.

5. James Martin, *Teleprocessing Network Organization* (Englewood Cliffs, N.J.: Prentice-Hall, 1970).

6. James Martin, *Systems Analysis for Data Transmission* (Englewood Cliffs, N.J.: Prentice-Hall, 1971).

7. Eugene V. Rostow, "The President's Task Force on Communications Policy," Staff Paper No. 1, Appendix G, p. 7. Washington, D.C., 1969.

SECTION  III

NEW TECHNOLOGY

Marconi's demonstration of the wireless on Salisbury Plain, England to skeptical officials of the Armed Services and Post Office. The Italian government had refused his offer of a demonstration and Marconi had come to England where he established the company that first developed the radio. Courtesy of The Marconi Company, Ltd., England.

13 THE RADIO SPECTRUM AND ITS PROBLEMS

Two different ways can be used to transmit information over cables and by radio transmission. In the former, different transmissions do not interfere with one another in an uncontrollable manner; they can travel over different cables and any number of cables can be used. In the latter, they do interfere and the radio frequencies available must be carefully allocated and rationed. The radio spectrum forms a natural resource that is in limited supply. As we move to high frequencies, however, the resource becomes more plentiful but has new problems, which we shall discuss.

In both cable and "wireless" systems there is a trend toward increasingly higher frequencies. It is this advance of the frontiers of technology into higher-frequency zones that gives us the increasing bandwidth and channel capacity discussed earlier.

THE SPECTRUM

Figure 13.1 shows the portion of the electromagnetic spectrum that is used for telecommunications today. Just as we may have a rainbow-colored spectrum of light by separating the visible frequencies into individual colors, so all the frequencies of physics may be laid out in a spectrum. Those shown in the figure are only a minute fraction of the frequencies known to physicists, who discuss rays of 10^{24} Hz and higher. The scale of frequencies in Fig. 13.1 is logarithmic. If we drew this diagram with a linear scale so that that part of the spectrum used today for commercial communication links occupied the width of this page, then the part of the known electromagnetic spectrum as yet unused would require a sheet of paper stretching from earth to a point twice the distance

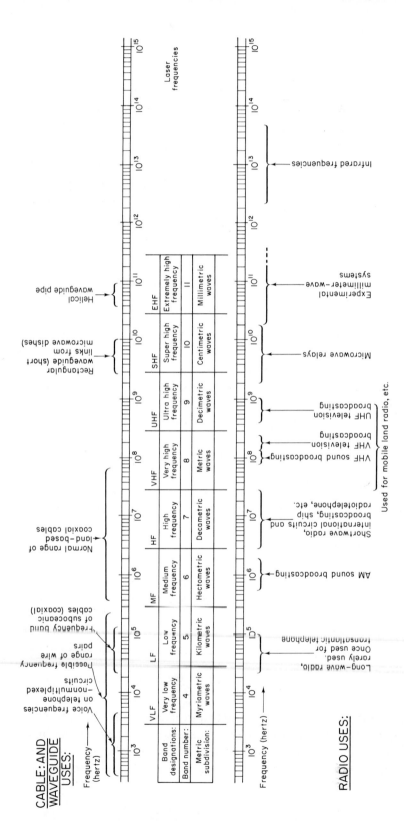

Figure 13.1. Telecommunication uses of the electromagnetic spectrum.

of the sun! Theoretically, the quantity of information the medium can carry is proportional to the distance in this scale.

Information-carrying capacity is approximately proportional to bandwidth (difference between highest and lowest frequency in a band of frequencies). The higher we go up the electromagnetic spectrum, the more plentiful is bandwidth. If we can extend the upper limits of radio usage from centimetric to millimetric waves (Fig. 13.1), we will increase the total bandwidth available by a factor of 10. These new frequencies have their own special problems, however, as will be discussed shortly.

It seems probable that the trend toward higher and higher frequencies for telecommunications will continue. We have no idea how to amplify or modulate frequencies much higher than those in use today. But, then, some of the systems in our laboratories now would have been entirely beyond the ken or even the credibility of engineers 20 years ago. By 1990 there will undoubtedly be new discoveries that would leave today's engineers incredulous.

RADIO
WAVES
This chapter discusses the use of radio frequencies; the next will discuss cables and waveguides.

Radio waves of different frequencies differ in their properties, which may be briefly summarized as follows:

1. Long-wave radio, 30 to 300 kHz

The original radio telegraphy frequencies. Once used for trans-Atlantic telephone. Little used today.

2. Medium-wave radio, 300 kHz to 3 MHz

Mainly used for the AM radio broadcasting frequencies. The familiar dial of the home AM radio tunes over a band from about 500 to 1500 kHz. The upper end of the band is used for mobile land transmitters and for amateur use.

3. Shortwave radio, 3 to 30 MHz

Used for shortwave radio broadcasting, international overseas ship radiotelephone, for aircraft, for much amateur use, and for mobile land transmitters.

Below 30 MHz, the spectrum is principally suited to very long distance transmission. Hence most of this region is used for international communication and broadcasting, as well as for maritime and aviation use. Government and nongovernment users share it almost equally. These frequencies are reflected by the ionosphere and therefore can bounce around the world. Being strongly scattered by the atmosphere, they are able to pass around large obstacles. Thus they are more easily used in car broadcast receivers than the higher frequencies. Tuning a domestic AM or shortwave radio set, the user can pick up stations from far away if the set is sensitive. This fact is especially true at night, for the ionosphere has better reflecting properties when not in sunlight. On the other hand, there is much atmospheric noise—noise largely absent at higher frequencies.

4. The VHF*band, 30 to 300 MHz

A large part of this band is also used for radio broadcasting; it is familiar to the domestic user as the FM band on his radio set. The user now must generally be within a hundred miles or so of the transmitter. The VHF band is also used for television broadcasting. Television needs a bandwidth many hundreds of times greater than sound broadcasting, and therefore the frequencies below this band could not be used for television.

The VHF band and the frequencies up to 1000 MHz are the most suitable for mobile communications on land. They are used for police, fire, and ambulance services, taxis, delivery fleets, car radiotelephones, personal paging systems, and so on. The demand for mobile radios is increasing rapidly, which means that this part of the spectrum is becoming overcrowded and there is an ever-growing shortage of these frequencies. Furthermore, technology has now brought us to a point where we would like to have a major increase in the numbers of portable transmitters and receivers. As Chapter 20 will show, an increase of one hundredfold could occur in the next twenty years if control of use of the spectrum permits. Fortunately, there *are* ways to achieve much more efficient allocation and use of the spectrum.

Atmospheric noise decreases with increasing operating frequency and

*VHF stands for Very High Frequency.
UHF stands for Ultrahigh Frequency.
SHF stands for Super High Frequency.
See Fig. 13.1.

eventually falls below the level of the thermal noise that is always present in electronic circuits. Radio circuits above 100 MHz are on a par with cable circuits. The high-quality FM radio can have an almost noise-free background (due also to the frequency modulation process which trades bandwidth for noise reduction).

Propagation is not affected by the ionosphere. It takes place in the lower atmosphere and becomes closer to a straight-line path as the frequency increases. The radiation is increasingly reflected by large objects. High buildings can cause "shadows" on television, and the helicopters over Manhattan interfere with FM radio and television in some locations.

5. The UHF band, 300 to 3000 MHz

The UHF band again has a capacity ten times greater than the lower band. A major block of it has been allocated for television. This band provides more channels than VHF, but these were allocated more recently and only a few are used as yet. Below about 1000 MHz, this band is used for mobile land radio, as is VHF, and shares its congestion problems.

Above 1000 MHz, the UHF band joins the SHF frequencies in being mainly useful for radar and navigational devices and the point-to-point communications.

6. The SHF band, 3 to 30 GHz

At these high frequencies, point-to-point transmission is generally necessary. The frequencies are thus too high for general mobile land radios. Atmospheric scattering is slight. Some circuits are used in which waves from powerful transmitters are scattered by the troposphere, but normally line-of-sight transmission is employed.

Frequencies in use today above 1GHz are referred to as *microwave*. Microwave radio is one of the chief means of transmitting long-distance telephone calls and television programs. As with coaxial cable, the high bandwidth of microwave channels can transmit many thousands of telephone calls, and many television programs, simultaneously. Microwave antennas, within line-of-sight of each other, are erected on tall towers and rooftops. These antennas and towers form chains across the country, each such location amplifying and retransmitting signals. They are usually spaced about 30 miles apart. If they were farther apart, the curvature of the earth and varying retractive index of the atmosphere would necessitate very high towers, absorption due to rain and snow would become severe, and large, expensive antennas would be needed.

Many cities throughout the world now have skylines that are dominated by a tower carrying microwave antennas. Tokyo has a tower like the Eiffel Tower but 40 ft higher. East Berlin has one 1185 ft high. One of London's most expensive dinners can be eaten in a revolving restaurant just above the Post Office tower microwave antennas, and Moscow possesses a tower that is 250 ft. higher than the Empire State Building.

Microwave relays today use frequencies from about 1 to 12 GHz. The higher the frequency, the easier it is to direct a narrow beam that does not interfere with other nearly transmitters using the same frequencies. This is one reason why the Datran Corporation chose 11 GHz for the frequency at which they will distribute their data signals in the cities. A beam of about 1° angle can be used, and the transmitter power for the city distribution would be low, again so as to lessen chances of interference. The bandwidth needed for this service is relatively low compared with television needs and compared with the high bandwidth available.

Different moisture and temperature layers can cause the beam to bend and vary in amplitude, just as we sometimes see light shimmering over a hot surface or causing mirror mirages along a road surface in the sun. Occasionally bending effects can cause fading. Rain can increase the attenuation slightly, especially at the higher microwave frequencies, and occasionally trouble is caused by reflection from unanticipated objects, such as helicopters or new skyscrapers in a city. To a limited extent, automatic compensation for changes in the radio attenuation is built into the repeaters.

Microwave frequencies are the ones used today by communication satellites.

7. Millimeter waves above 10 GHz

The new frontier of radio is in millimeter waves. The bandwidth is enormous and so is the payoff in discovering how to use it.

Waves of these frequencies suffer severe attenuation in rain and snow. Below 6 GHz, the effects of rain play no part in the spacing of microwave relay towers. Above 10 GHz, the effects become severe. At 20 GHz, the spacing cannot be more than a few miles. Millimeter-wave systems will probably be used in cities, with relay-tower spacing of less than half a mile, as we will discuss shortly.

The higher the frequency, the smaller the antenna needed to make a narrow-angle beam. Small antennas on tall posts in city streets can relay millimeter-wave signals of extremely high bandwidth. Because of the short range and narrow angle beams, different city pathways can be prevented from interfering with one another.

8. Infrared and optical transmission

Systems are coming into use that employ light beams to carry signals. Some systems use frequencies lower than those of light in the infrared zone. Frequencies in the range from 10^{13} to 10^{15} might be employed.

At these frequencies the beams travel in straight lines and can be sharply focused with lenses. Consequently, there is no problem with interference. On the contrary, there may be a problem with vibration of the light source because the beam is so accurately focused. No license is needed from the FCC to operate at these frequencies. The big snag with these frequencies is that they are heavily absorbed in thick fog, as we can see with the unaided eye. The signal can also be blotted out with very intense rain. Because of these effects of bad weather, optical and infrared transmission is normally limited to short distance links.

We shall discuss both these frequencies and millimeter wave frequencies shortly.

Figure 13.2 shows the antennas used for different radio frequencies.

ALLOCATION OF THE RADIO SPECTRUM There are both international agreements and more detailed national regulations on the usage of the radio spectrum. In some countries the national frequency allocations are "classified". In the United States they are not, and the U.S. National Table of Frequency Allocations extends from 10 kHz to 90 GHz. It has hundreds of bands of frequencies allocated to different uses—government, military, and civilian. Bands allocated to one service are often separated by bands allocated to another. Some bands are allocated to two or more services on a shared basis.

Figure 13.3 summarizes the way the bands are allocated. The frequencies from 30 to 1000 MHz are the most congested, these being the frequencies needed for television and radio broadcasting and for mobile radio services. The FCC [1] has described the situation in the land mobile services as being one of "congestion" (1958), "extreme congestion" (1962), and "acute frequency shortage" (1964). By now the frequency

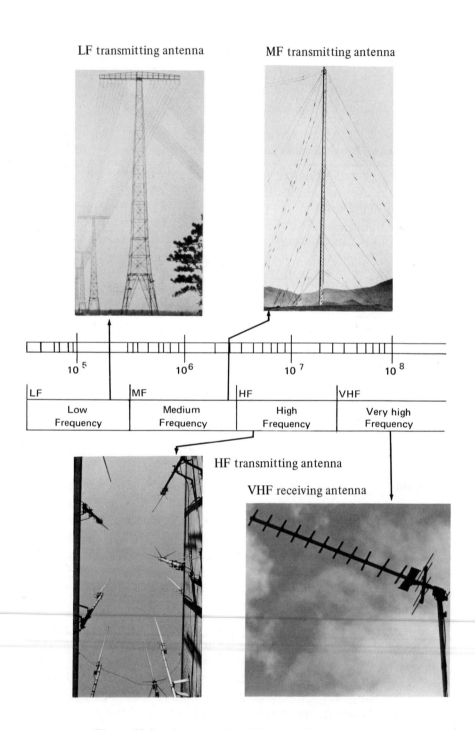

Figure 13.2. Antennas for different radio frequencies.

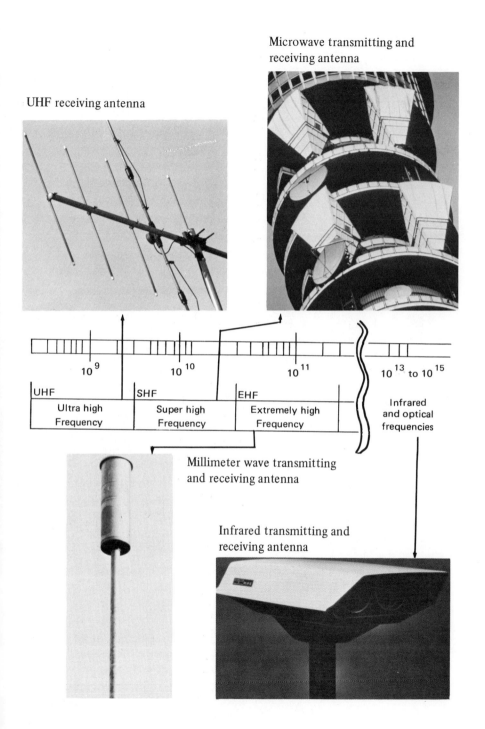

Microwave transmitting and
receiving antenna

UHF receiving antenna

10^9 10^{10} 10^{11} 10^{13} to 10^{15}

UHF | SHF | EHF

Ultra high
Frequency

Super high
Frequency

Extremely high
Frequency

Infrared
and optical
frequencies

Millimeter wave transmitting
and receiving antenna

Infrared transmitting and
receiving antenna

Figure 13.2

177

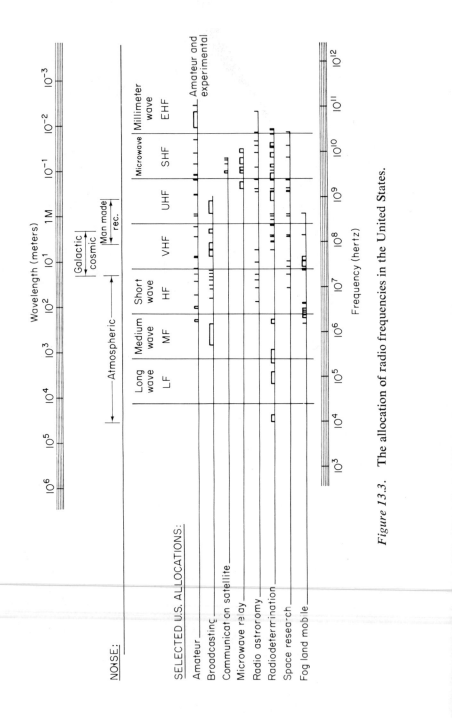

Figure 13.3. The allocation of radio frequencies in the United States.

178

shortage is definitely impeding progress in developing mobile radio services (discussed in Chapter 20).

The congested area is shown in more detail in Fig. 13.4; this time the international frequency allocation chart is given. Usage of frequencies is more internationally uniform below 30 MHz because it is in these lower frequencies that worldwide transmission is possible.

WAYS TO ALLEVIATE THE FREQUENCY SHORTAGE   Fortunately, there are many ways in which the spectrum congestion can be relieved. These methods involve not merely a better sharing of the spectrum but better uses of today's technology. They cannot be achieved without national and, in some cases, international cooperation and control. In the United States, it seems highly desirable to establish an agency that continually reviews and reallocates the nation's electromagnetic resources. New problems will arise with the efficient use of satellites, which must be prevented from interfering with one another and with other land resources. It is necessary to consider these problems along with other spectrum problems.

Let us summarize the steps that can be taken to alleviate the spectrum problems. It is important that action be taken on some of these points as quickly as possible because of the great speed with which this industry is evolving and because of the greatly changed demands that computer technology and large-scale integration circuitry are bringing.

1. Better allocation of the frequencies

Some services for which blocks of frequencies are allocated are underutilized while others are bursting at the seams. Bands allocated for forest conservation, forest products, and highway maintenance are underutilized, for example, whereas those for business radio and radio telephone are so crowded that they are holding up progress of a nature which could have a valuable effect on the economy. The wide band allocated for 70-UHF television channels is little utilized today, and the unused portion could carry many thousands of radio telephone channels. Some military allocations are also underutilized. Some parts of the spectrum are not allocated at all because of possible future needs.

Spectrum allocations must be changed as the needs and technology change. The spectrum should be more finely apportioned, in smaller blocks, and the allocation should be much more dynamic.

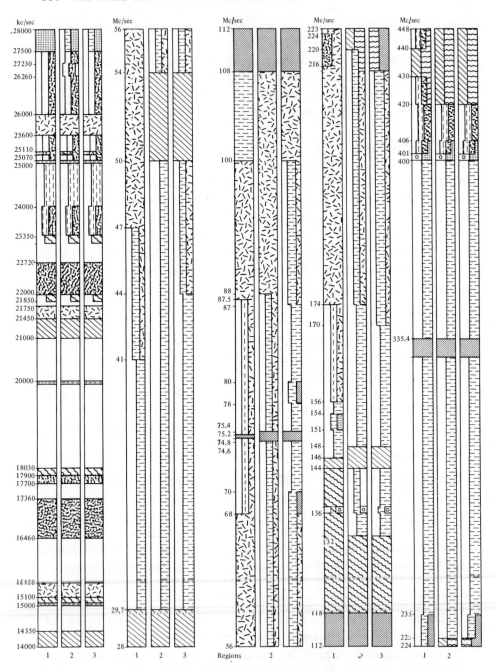

Figure 13.4. International frequency allocations, showing the most crowded part of the spectrum. *By permission of the Radio Corporation of America.*

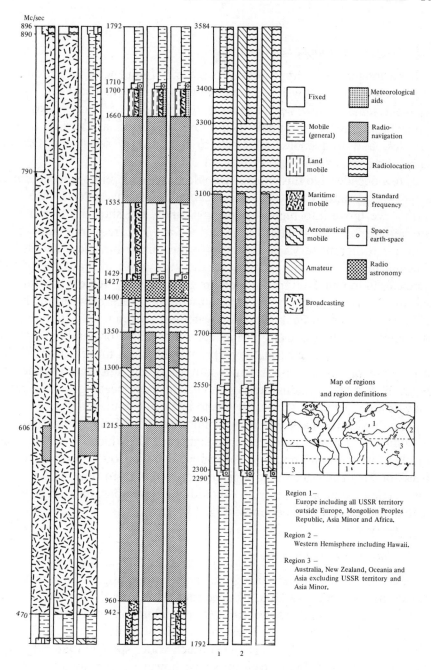

Figure 13.4. Cont.

2. Frequencies should be shared

It no longer seems appropriate to allocate frequencies in nationwide blocks. At the time of writing, the band allocated to the petroleum industry is the same in New York City as in the oil fields of Texas. Taxis have the same number of channels in the Oregon forests as they do in Los Angeles, and frequencies allocated to the forestry service in Oregon cannot be used, except for forestry, in Los Angeles.

A few decades ago this situation would have made more sense because longer wavelengths were used, which caused long-range interference. The farther we move into higher-frequency bands, the shorter the range of interference and the greater the need for *localized* frequency allocation.

Again, different services do not have the same utilization at different times of the day or different times of the year. Most business use of frequencies in mobile land transmitters, paging schemes, vehicle control and scheduling, radiotelephones, and so on will take place between the hours of 8 A.M. and 6 P.M. Peak television watching is from 6 P.M. to midnight. Automatic machines, transmitting computer data, or sending mail in facsimile form could operate at night. Would it not make sense to allocate some of the television bandwidth to other functions outside the hours of 6 P.M. to midnight? Again certain users needing mobile radio may be able to confine their activities to just a few hours a day.

The President's Task Force on Communications Policy commissioned a study on spectrum management techniques [2] which advocated that market forces should apply to spectrum allocation on a basis of "packages" of frequencies covering a given area and a given time. "TAS packages" (time, area, and spectrum) should be used rather than the National Table of Frequency Allocations (Fig. 13.3).

3. Narrower transmitter bands

Much of the equipment using the spectrum is designed in a way that wastes its frequencies—after all, spectrum use is free. In particular, most mobile voice transmitters use 100 kHz per telephone channel—the same telephone channel which receives only 4 kHz of equivalent bandwidth on the Bell System. Radio systems need much wider spacing than cable systems to separate different transmissions; however, with today's state-of-the-art, they do not need 100 kHz. 20 to 25 kHz would be adequate. In the Los Angeles area, for example, consideration is being given to dropping the channel spacing for marine radio from 50 to 25 kHz, which will double the available number of channels. A drop from 100 to 25 kHz would quadruple the number of mobile land radio channels, with a fairly small increase in transmitter and receiver cost.

4. Party lines and dynamic channel allocation

With mobile radio transmission, one frequency in one location can be regarded as a channel. Several users may share this channel, just as they can share a telephone party line. If one person is using the channel, other persons wanting to use it must wait until he has finished. Although party lines are becoming less common on the telephone network, the number is increasing in radio because of the shortage of frequencies. In many cases, the number of users per channel is becoming inconveniently high so that long waits are encountered. However, the spectrum utilization is increased greatly.

The waiting time for a given channel utilization can be reduced if the user is permitted to use one of *several different* channels rather than one fixed channel. If he is driving up to a set of toll booths on a highway, he can select whichever booth is free, for example. The probability of his having to wait is lower than if he always had to use the leftmost booth. In the jargon of queuing theory, it would be a "multiserver" rather than a "single server" queuing situation. Figure 13.5 shows how the waiting time for queues varies with the numbers of channels (or servers). These curves assume an exponential distribution of the channel-holding times and calls arriving at random—that is, a Poisson distribution of number of calls per unit time. These conditions approximate the situation on typical mobile radio channels. They also assume that callers wait until a channel becomes free. (These curves and related queuing theory are discussed in James Martin, *Systems Analysis for Data Transmission,* Prentice-Hall, 1971.)

It will be seen that the queuing time is less for the multichannel queues. With a channel utilization of, say, 0.5, there is a substantial waiting time if only one channel is used. It is less if two channels are used. If ten are used, the mean waiting time is very low indeed.

To take advantage of this a mobile radio transmitter and receiver could be built to operate on several different channels. When it comes into operation, it searches for a free channel and then uses it. This process adds to the cost of the transmitter and receiver, but, once again, in an age

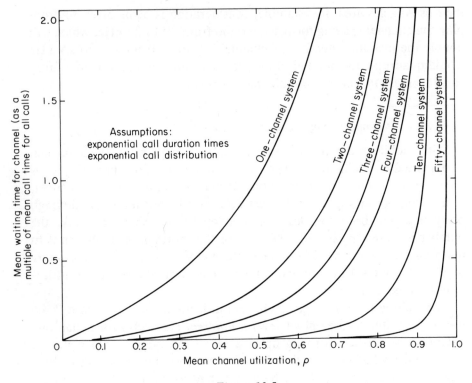

Figure 13.5

of large-scale integration, circuit logic for the searching operation will be inexpensive.

5. Low-power transmitters

A television transmitter has high power and beams its signal over distances of 100 miles or so. Nothing else over a wide area can use its frequencies. A taxi fleet operating within a city does not need such a powerful transmitter. Separate taxi fleets in nearby cities could use the same frequencies if the transmitters employed were suitably low in power.

In any area over which continuous coverage is needed, it is often desirable to reuse the frequencies for different transmissions in different locations. A vehicle fleet in one area should not be prevented from reusing the same frequencies as a vehicle fleet in another area. How far apart the same-frequency zones are depends on the design of the transmitter and receiver. With modulation techniques designed to give interference-free reception, the same-frequency zones can be separated by one zone of a different frequency. Consequently, in order to give complete coverage of a wide area, four different frequencies are needed, as shown in Fig. 13.6. With less-expensive design, separation by two or more zones is needed. Two-zone separation needs nine different frequencies, as shown in Fig. 13.7. Separation by N zones needs $(N + 1)^2$ different frequencies.

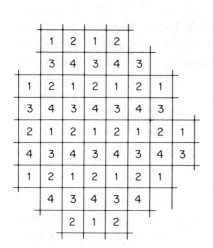

Figure 13.6. Interlacing of four frequencies separates zones using the same frequency by one zone or more.

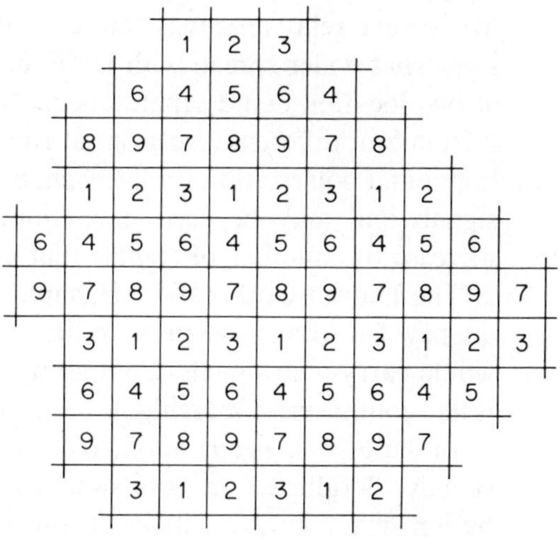

Figure 13.7. Interlacing of nine frequencies separates zones using the same frequency by two zones or more.

In order to serve as large a number of mobile radios as possible, the zones should clearly be small. In other words, the transmitters should be low in power. It will be necessary in future systems (see Chapter 20) for vehicles to travel from zones of one frequency band to zones of another and for calls to and from them to be appropriately and automatically switched.

If mobile transmitters cover a distance of 4 miles and same-frequency zones are separated by one zone, then each frequency used must be unique within an area of 64 square miles. For television transmitters covering a distance of up to 100 miles, the equivalent area will be 10,000 square miles. The frequencies used by one television channel could be used for over 200 mobile radio channels with a channel spacing of 25 kHz. But because of its wide area coverage, one television broadcast uses frequencies that could be used by over 100,000 mobile radio channels with appropriately low power transmitters.

6. Directional antennas and polarization

Transmitting antennas can be built to transmit in one particular direction, and receivers can be built to receive from that direction only. The higher the transmission frequency, the narrower the transmitted cone. Microwave relay antennas can use a very narrow beam. The beam has a somewhat wider spread with UHF and VHF bands. However, a receiver at one location could separate signals of the same frequency beamed to it from four different directions at right angles to each other. Vertical and horizontal polarization similarly enables the separation of same-frequency signals and may be used in conjunction with directional antennas to increase the number of signals that can be distributed in a city.

The Datran Corporation (Chapter 9) proposed 11 GHz as their frequency for transmission in cities. Low-powered microwave antennas would carry such signals. Comparing this with today's television antennas is like comparing a searchlight beam with a floodlight.

In the diverse needs of our cities for the distribution of many types of signals "floodlight" transmission is far too wasteful. Much radio distribution in the future will be in the form of criss-crossing "searchlight" beams spanning the rooftops; the narrower the beams the better. To achieve these narrow beams higher frequencies are needed and hence the 1970s will see radio techniques pushing up beyond the microwave frequencies into millimeter waves, infrared and optical transmission.

7. Maximum use of cables rather than radio

Television distribution of the future may use the "searchlight beam" to carry signals to the head of cable TV systems. The city antennas may be similar to those in Fig. 9.5. If these steps are taken, the large bandwidth reserved for television today could be freed, at least in certain cities, for mobile radio services. As television changes from "wireless" to cable distribution, large blocks of frequencies could become reallocatable. This much-advocated viewpoint meets staunch opposition from the broadcasting companies, however, so many important uses can now be projected for mobile radio that it would be of great value to remove from the VHF and UHF radio bands any signals that can be distributed by cable or microwave radio.

Some authorities have advocated that the UHF television band be reallocated partially; others favor reallocating completely, leaving only VHF and cable TV. Nevertheless, it seems unnecessary to reallocate it nationwide. Some areas of the country may not have cable TV for a long time. The very high capacity land lines which we will discuss in the next chapter are being used increasingly for television distribution. These lines could carry all types of signals to TV cable heads or to local broadcast transmitters which do not blanket-out such large areas as do most of today's transmitters. The flooding of large areas with very high transmission prevents alternate use of the frequencies. Broadcasting of the future should consist of a judicious combination of cable distribution and small localized transmitters, leaving the maximum "time-area-spectrum packages" for other use.

8. Use of higher frequencies—millimeter waves

If systems can be designed to utilize millimeter waves, the bandwidth available for radio transmission will become several times the total of that used today. Millimeter waves permit highly directional antennas and will be used mainly for "searchlight beam" distribution of signals within cities. They may also come into use for communication via satellites.

The main problem, apart from the design of circuitry for such high frequencies, is that these frequencies are absorbed heavily by rain and snow. Certain frequencies are absorbed heavily by the atmosphere itself. The weather, in fact, has recently assumed such importance in telecommunications that articles on rainfall and other aspects of meteorology have started to appear in the austerely technical *Bell System Technical Journal*.

Figure 13.8 shows the attenuation that is caused at different frequencies by heavy rain. Rain measurements made in the United States were somewhat lower than the solid line in this diagram. If the fading caused by rain is permitted to be no more than 40 db, this amount sets a limit on how long the radio path can be. Figure 13.9 is a diagram from the Bell Laboratories showing the distance between repeaters that could be used at different frequencies. The spacing is about 6 miles at 11 GHz. (This figure is the city distribution frequency proposed for the Datran System.) At higher frequencies, it drops to 2 or 3 miles.

The amount of heavy rain varies widely from location to location. Figure 13.10 shows the number of minutes per year that the attenuation on a 30-GHz radio path exceeds certain limits. The result is not directly

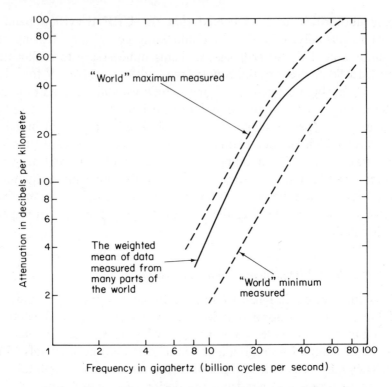

Figure 13.8. The attenuation caused by rain at the rate of 100 mm (4 inches) per hour measured in many parts of the world. *Redrawn from reference 6, by permission of AT &T.*

related to the amount of time it rains because it depends upon the intensity of the rain. England, where one rarely goes out without a raincoat, does better than Miami, Florida, because when it does rain in Miami it rains cats and dogs. Very heavy rain is often isolated geographically into localized storms. The most intense storms are, in fact, the most limited in area, which raises the possibility of automatically switching transmission to alternate paths when storms occur.

The advantages of transmission above 10 GHz are the very high bandwidths available and the fact that the radio beams are highly directional; thus an urban area can have many crisscrossing beams (like searchlights) that do not interfere with one another. These advantages, together with the shortage of spectrum space elsewhere, will override the disadvantages of attenuation.

The high attenuation, however, means that short hops with many re-

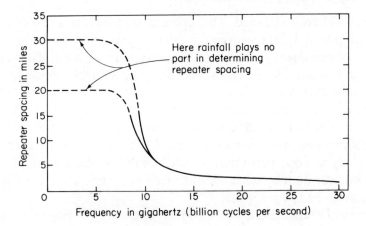

Figure 13.9. Repeater spacing as determined by a rain rate of 100 mm (4 inches) per hour. This assumes uniform rain over the entire path and a 40-dB fade margin. *Redrawn from reference 3, by permission of AT &T,*

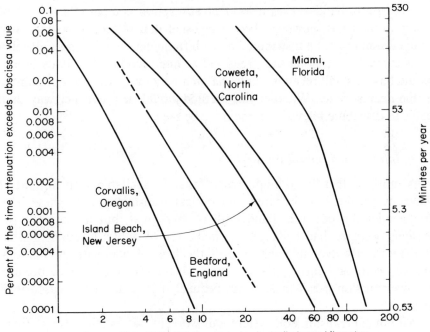

Figure 13.10. The number of minutes per year that a given level of attenuation at 30 GHz is exceeded for parts of the United States and England. *Reproduced from reference 7, with permission.*

peaters will have to be used. The repeater, therefore, must be of low cost and high reliability. It must require little or no manual attention or maintenance. Consequently, solid state circuitry must be used. The repeaters may stand on poles in the city streets; in this case, their appearance must not be unattractive.

Figure 13.11 shows a city repeater designed at the Bell Laboratories [4]. It operated at 11 GHz and met the preceding requirements; it received and transmitted over a distance of 1½ miles. The repeater proved satisfactory in cost, performance, and reliability. Experimental systems have also been built by other organizations, some operating at frequencies higher than the foregoing one.

The main characteristics of radio relay systems operating at frequencies above 10 GHz, as seen at the Bell Laboratories, are summarized in Fig. 13.12.

Going still higher in frequency, the absorption caused by the atmosphere itself becomes significant (see Fig. 13.13). At 60 GHz, the first O_2-absorption band causes a level of absorption comparable to that of heavy rain. It has, however, been suggested that wideband signals such as television might be distributed at such frequencies by using small relay units every 300 ft along the street and at intersections. The householder would have a small receiving antenna on a windowsill or rooftop, pointed at the nearest relay. Whether this method would be a cheaper way than cable to distribute television remains to be seen.

9. *Infrared and optical transmission*

A more perfect "searchlight beam" than is available with millimeter waves is provided by infrared and optical transmission. At these high frequencies the potential bandwidth is very great, but as of 1971 the systems designed had been able to make use of only a fraction of what is theoretically available, because of the lack of adequate modulators. As we mentioned earlier, there is no problem with interference because the transmission uses a pencil-thin beam that is accurately focused with lenses.

An optical transmission system can work in two possible ways. First, a laser can be used and the waves transmitted can be coherent—that is, the same phase relationship exists throughout the beam. The receiver could then use heterodyne detection, as in radio reception, in which a fixed frequency is added to the received signal to produce fluctuations (or beats) or a frequency equal to the difference between the two signals.

The second approach is cruder. It uses fluctuating noncoherent light (or infrared signals) and detection is accomplished by photon-counting schemes—that is, direct detection of the incoming light with some form of photodiode. Coherent transmission and heterodyne detection was expected to be the better scheme. Background noise is not amplified with the signal and a larger information bandwidth is possible. However, noncoherent transmission has, so far, proved to be less inexpensive and it is less perturbed by atmospheric path fluctuations.

The success of noncoherent optical transmission has largely been due to a device called the *light-emitting diode* (referred to as *LED*). This is a small semiconductor diode that emits radiation when a current is passed through it. The frequency of the emission depends upon the material of which the diode is made. It is desirable to select a material that emits at a frequency to which a radiation detector is most sensitive. Galium arsenide is commonly used as this emits at the sensitivity peak of a silicon detector. This tiny diode emits light approximately proportional in intensity to the current used.

The information rate that can be transmitted with a light-emitting diode is limited by the fact that the device takes a certain time to turn on. Typical diodes in use today take from 2 nanoseconds (2×10^{-9} second) to 100 nanoseconds to rise to full emitting power. Such devices have been used to transmit two TV channels or equivalent data rates.

Figure 13.14 shows an inexpensive optical data transmission unit in operation. This device, the CTS 1815 OPTRAN transceiver can transmit and receive up to 250,000 bits per second. It is typically sited on the roof of a building, where it is inconspicuous because of its small size ($19'' \times 12'' \times 5''$). It can transmit more than a mile in clear weather, however, it is more likely to be used for distances of 500 to 300 ft, depending on local weather conditions. The Datran Company has considered using this transceiver for parts of their city distribution network. It is competitive in cost with their microwave distribution system described in Chapter 9. The severe problem, however, is the loss of the transmission path in thick fog or very intense rain.

IBM used optical transmission at Expo '67 in Montreal. Their links are shown in Fig. 13.15. One carried two television channels a distance of almost two miles. The other was a data link carrying 1.3 million bits per second over a distance of about half a mile.

Typical availability figures on optical links have been about 98% because of bad weather. Optical transmission engineers today can be found

Figure 13.11. An experimental repeater for relaying radio signals of frequencies above 11GHz down city streets. Built at the Bell Laboratories.

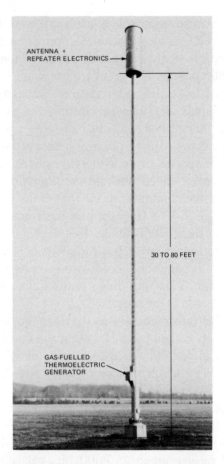

Receptors such as this may be sited every one to two miles.

Photographs courtesy of Bell Telephone Laboratories.

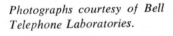

A front view without the weather cover or inner absorbing liner.

A side view without the outer skin, showing the internal structure.

The base of the column.

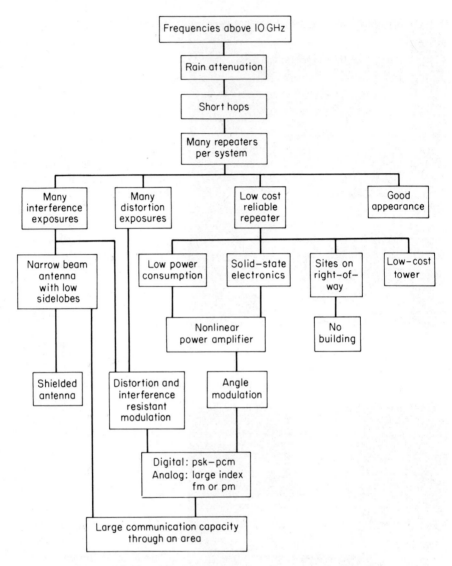

Figure 13.12. Characteristics of radio systems operating at frequencies above 10 GHz. *Redrawn with permission from the Bell System Technical Journal (4).*

eyeing the weather like yachtsmen. In cities that have little fog, the figure can be much better, and we may well see a mesh of optical links being built, spanning the rooftops. Areas that are frequently fog-bound will probably use some other means of distributing signals. For certain types of transmission, such as the sending of batches of computer data, or facsimile mail, the periodic ontages of an optical link may not matter. It is interesting that the availability figure for leased telephone lines in the United States is typically only about 98% also.

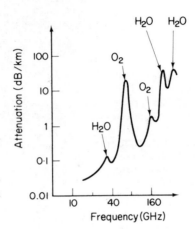

Figure 13.13. The absorption of millimeter waves in the atmosphere (sea level) with no rain.

SPECTRUM ENGINEERING
One of the much-used phrases of the next decade will probably be *spectrum engineering*. This term refers to the "next generation" of spectrum management techniques in which the spectrum acreage is allocated much more flexibly than with today's National Table of Frequency Allocations. The allocation of blocks of frequency, time, and area will be tailored to particular situations.

Much of today's equipment and many of today's operating practices result in severe spectrum wastage. Spectrum engineering will tighten the efficiency of spectrum usage with all methods that prove economical, including those just discussed. Today's users commonly employ broader bandwidths than are necessary, as well as higher power, less-directional antennas, less-effective modulation techniques, less than optimum antenna locations, and no polarization. Large acreages of spectrum are today allocated to the distribution of signals that would be better sent by cable. Spectrum engineering will attempt to apply standards and operating criteria that would reduce the wastage.

A detailed study of spectrum engineering requirements was carried out by the IEE/EIA Joint Technical Advisory Committee (JTAC) [8]. According to the committee, two characteristics should dominate future spectrum engineering.

Figure 13.14. Optical and infrared transmission may form the basis of inexpensive signal distribution schemes in the cities. Part (a) is the Optran transceiver that transmits and receives, using noncoherent optical transmission, at a rate of 250,000 bits per second.

(a)

(b)

Part (b) is an installation of the Optran device under test by the Datran Company. The unit on the left is transmitting to a similar device on a roof hidden in the mist behind the church steeple, half a mile away. *Photograph courtesy of Datran Company.*

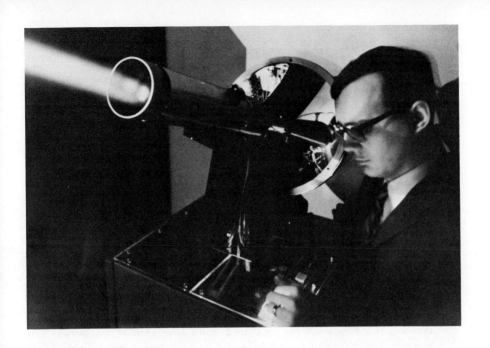

MONTREAL STOCK EXCHANGE

Infrared
Transmitter

Figure 13.15. IBM used infrared transmission at Expo '67 to transmit TV images and data, as shown. Stock market quotations were transmitted from the Montreal Stock Exchange a distance of 2 miles and a 2250 graphic display was connected to a 360 computer half a mile away.

Distance
2 Miles

Data
and
Video

Infrared
Receiver

Infrared
Transmitter
and
Receiver

TV
Barrage

2916 – 111
Line
Adapter

Distance
$\frac{1}{2}$ Mile

TV
Monitor

360/50
Computer

CANADIAN PAVILION

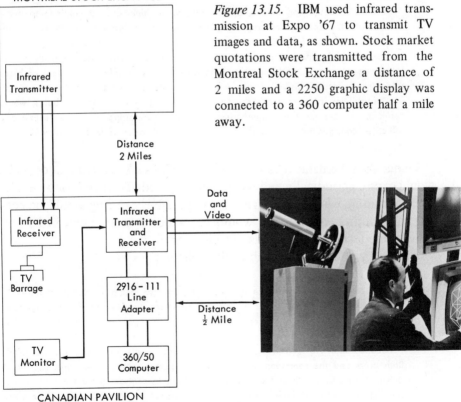

197

First, the system must be evolutionary. The present system's functioning is understood, in varying degrees, by many thousands of people—millions, if licensees' operators and Amateur and Citizens Radio Services licensees are included. Their loyal, if often exasperated and occasionally imperfect, support of the system is in a way the bedrock of the nation's spectrum usage structure. Changes in the system must be well thought out and amply foreshadowed, or the continued cooperation of those concerned will simply not be possible. Also, recovery of capital investment in equipment must be considered in planning for changes. But none of this means that evolution must continue at the snail's pace of the past.

The JTAC anticipates that the system must continue to evolve. The "next-generation" system envisaged by the JTAC cannot become operational in full before the early 1970s (though some components of it could and should be advantageously put to work at once). It will have capabilities far surpassing those of the present system, and its basic philosophy is much more flexible and effective; but it will incorporate old concepts of proved continuing usefulness and will still be a clearly recognizable descendant of previous systems. However, the system must continue to evolve beyond that version into still different forms in order to function effectively under the radically changing conditions which are anticipated in the later years of the century. It is important that this evolutionary development be much more dynamically responsive than it has been.

Second, spectrum engineering and management thinking must continue to move away from the concept of controlling spectrum usage through simple but rather restrictive and rigid administrative rules. The movement must be in the direction of increasingly individualized technical assessment of applications under explicitly formulated priority criteria, and under a reduced and more flexible employment of block allocation concepts. This will require much heavier use of analytical and data processing capabilities in the nation's spectrum engineering than at present; but it would result in stronger spectrum engineering capable of supporting the more flexible and effective management needed if fuller spectrum utilization is to be achieved.

Under the scheme advocated by the JTAC, user's spectrum allocations would expire periodically and have to be renewed. The needs of would-be users and users whose authorization to operate is expiring would be considered together. An attempt would be made to fit all users in by employing the latest developments in technology.

Three key points are implicit in this scheme:

First, objective measures are necessary of the use being made of channels. It may be surprising but it is fair to say that spectrum engineering and management, lacking these data, are at present like a merchant or supply officer trying to function without an inventory record.

Second, objective measures to assess applicants' needs for service are necessary. Both users and the spectrum engineering/management system need this. Lacking it, the user is in the position of a householder about to contract for a fixed amount of fuel for the winter without knowing anything about his locality's winters or about the quality of the insulation of his house. The spectrum engineering/management system now has no easy way of telling whether an applicant's statement of his needs is well-founded or not.

Third, applications for renewal of authorizations must be weighed in the same manner as those for new authorizations. If and when priorities are applied to applications, it must be possible to replace one operation with another of higher priority, even if the one is of long standing and the other is new, at the expiration of the older user's authorization [8].

Under the JTAC scheme, computers would maintain a detailed data base of current spectrum users and their equipment characteristics. They would also keep a data base of available spectrum acreage, and the computers would be used to assist in the selection of frequency assignments. The usage of assigned frequencies would be carefully monitored.

CHARGING FOR SPECTRUM USAGE   The merits of better spectrum engineering are indisputable. However, it can be argued that rather than having an elaborate government agency for the allocation of user priorities, market mechanisms should be allowed to play a part. Spectrum acreage, like land acreage, this argument says, should be bought and sold, the price reflecting the relative strengths of supply and demand in different parts of the spectrum.

Where spectrum space is short, the price will rise. It is unlikely that there would be large, unused tracts of prime acreage like the present UHF television channels. If the cost of the large bandwidth of a television channel becomes too high, the result will be a big incentive for cable TV.

A fairly difficult technical problem exists in how to define markettable "plots" in such a way that their owner cannot interfere with other users. The time-area-spectrum (TAS) packages mentioned earlier seem suitable at VHF and lower frequencies. At higher frequencies, however, directivity of antennas, polarization, reflection off buildings, and other factors make the demarkation of electromagnetic "plots" more difficult.

If such problems can be solved, perhaps the main objection to a market mechanism is the conflict between what is profitable for businessmen in radio usage and what is socially desirable. The maximizing of advertisement revenue, for example, makes for poor television. Radio astronomers would probably suffer if there were a free market in spectrum space. Educational uses of the media would be held back if the spectrum price was high.

The best compromise, perhaps, would be to make only certain parts of the spectrum available for "marketing."

REFERENCES

1. Eugene V. Rostow, "The Use of Management of the Electromagnetic Spectrum," Staff Paper No. 7, the President's Task Force on Telecommunications Policy, Washington, D.C., 1964.
2. "Electromagnetic Spectrum Management: Alternatives and Experiments," the TEMPO Study, General Electric Corporation 1969.
3. L. C. Tillotson, "Use of Frequencies above 10 GHz for Common Carrier Application," *Bell System Tech. J.*, July–August 1969.
4. C. L. Ruthroff, T. L. Osborne, and W. F. Bodtmann, "Short-Hop Radio System Experiment," *Bell System Tech. J.*, July–August 1969.
5. A. B. Crawford and R. H. Turrin, "A Packaged Antenna for Short-Hop Microwave Radio Systems," *Bell System Tech. J.*, July–August 1969.
6. R. A. Semplak and R. H. Turrin, "Some Measurements of Attenuation by Rainfall," *Bell System Tech. J.*, July–August 1969.
7. D. C. Hogg, "Millimeter-Wave Communication Through the Atmosphere," Science No. 3810, January 5th, 1968.
8. "Spectrum Engineering—The Key to Progress," Joint Technical Advisory Committee of the IEE/EIA, report for the President's Task Force on Communications Policy, Washington, D.C., 1969.

Three-stage de Forest audion amplifier of the type first built in 1912 by the Federal Telegraph Company. This is the earliest known commercial cascade audio-frequency amplifier. It had a gain of 120. Courtesy ITT.

14 ULTRAFAST CHANNELS

In view of the limited availability of radio spectrum space, any signal that can economically travel by cable rather than by radio (below 60 GHz) should be made to do so. This fact is especially true in the VHF and UHF regions, which are needed for mobile transceivers. Television sets can receive their signals via cable; but a taxi, delivery truck, or automobile cannot drag an umbilical cord around the street after it.

The information-carrying capacity of cables has been steadily increasing ever since Gauss and Weber first strung wire over the roofs of Gottingen in 1834. The highest-capacity cables in public use today are the coaxial cables used by the telephone companies for their main telecommunication highways. The next step up the scale is the helical waveguide, which AT & T is about to start laying for Bell System use. Then comes the prospect of laser channels. This chapter discusses these very high capacity facilities.

As cables of larger capacity are employed, so increasing numbers of voice, television, or other signals can be sent over one cable, a factor that will lower the cost per channel-mile. As mentioned in Chapter 3, the investment cost of adding a voice channel-mile to the Bell System today averages $11, and this figure is expected to drop to $1.40 by 1979; in the following decade, some authorities expect it to drop again to one tenth of the 1979 figure [1].

Two new types of cables for high-volume links are about to be laid in the Bell System—the CLOAX coaxial cable and the helical waveguide.

COAXIAL
CABLES

The common-carrier cables at the end of World War I consisted of many twisted pairs of wires grouped together in protective sheaths. These cables presented problems because of cross talk between separate pairs of wires and because of high signal losses. They could not carry a high frequency and so were unable to transmit broadband signals. There was no need to transmit broadband signals in the late 1940s, but the dawn of television was on the horizon. By 1950 several thousands of miles of coaxial cable were in use for telephone transmission in the United States.

As the signal frequency is increased on a conducting wire the current tends to flow increasingly on the outside surface of the wire. The current uses an increasingly small cross section of the wire, and thus the effective resistance of the wire increases. This is called the *skin effect*. Furthermore, at higher frequencies, an increasing amount of energy is lost by radiation from the wire. Nevertheless, it is desirable to transmit at as high a frequency as possible so that as many separate signals as possible can be sent over the same cable. The skin effect limits the upper frequencies. Crosstalk also imposes limitations.

A coaxial cable can transmit much higher frequencies than a wire pair. It consists of a hollow copper cylinder, or other cylindrical conductor, surrounding a single-wire conductor having a common axis (hence coaxial). The space between the cylindrical shell and the inner conductor is filled with an insulator, which may be plastic or mostly air, with supports separating the shell and the inner conductor every inch or so. A coaxial cable is shown in Fig. 14.1.

Several coaxial tubes are normally bound together in one large cable (see Fig. 14.2). At higher frequencies, there is virtually no cross talk between the separate coaxial cables in such a link because the current now

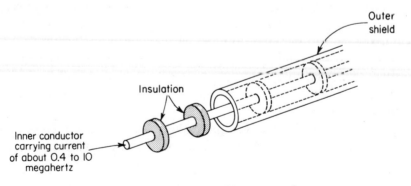

Outer shield

Insulation

Inner conductor carrying current of about 0.4 to 10 megahertz

Figure 14.1.   Coaxial cable construction.

Figure 14.2. A cable with 20 coaxial units, which can carry 32,000 telephone calls at once, or the equivalent other forms of signal. *Courtesy AT & T.*

tends to flow on the inside of the outer shell. Because of this shielding from noise and cross talk, the signal strength can be allowed to fall to a lower level before amplification.

The coaxial cables of the late 1950s used 0.375-in. tubes with a top frequency of about 3 MHz. A cable typically carried four such tubes, occasionally six. The engineers of the day used to question how anybody could use more bandwidth than that!

Before long, however, the top frequency was raised from 3 to 9 MHz, and the number of one-way voice signals that a single tube could carry rose from 600 to 1800. Now the top frequency of this cable is about 20 MHz and one tube can carry 3600 voice signals. Two tubes are needed for 3600 two-way telephone conversations. The cable shown in Fig. 14.2 has 20 such tubes; two of these are reserve links in case of failure; the remaining 18 can carry $9 \times 3600 = 32,400$ two-way conversations.

THE CLOAX CABLE The process of raising the frequency and increasing the capacity continues. Engineers are now considering tripling the frequency of today's cables for analog signals. The CLOAX cable is the Bell System's latest design. It uses a new type of fabrication with a corrugated copper-steel laminate as the outer conductor. (CLOAX stands for Corrugated-Laminated COAxial.) It is more economical to make and is both stranger and better electrically than its predecessors. The CLOAX cable will carry more than 80,000 simultaneous voice calls or their equivalent.

DIGITAL VERSUS ANALOG SIGNALS   On twisted wire, coaxial cable, or other transmission media, much higher frequencies can be used if a digital pulse stream rather than an analog signal is transmitted. The reason is that detection of the simple presence or absence of pulses, in spite of distortion and noise, is much simpler than the careful preservation of the waveform needed for an analog signal. The pulses of data are *reconstructed* at each repeater. If a 1 or 0 bit can be detected correctly, a new bit can be transmitted by the repeater station. This regeneration is the most valuable asset of digital transmission since it can prevent the accumulation of distortion that is inevitable in analog amplification. With analog repeaters, the noise and distortion are amplified along with the signal.

In Chapter 6 we mentioned that single twisted-wire pairs carry 1.5 million bits per second in the Bell T1 carrier system. This amount is about 20 times higher than the bit rate that could be transmitted over the wire pair with modulation of an analog signal. Similarly, if the coaxial cable is used for digital transmission, frequencies about 20 times higher than those used for analog signals can be employed. Consequently, *pulse code modulation* becomes an economic proposition on coaxial cables, a fact we shall discuss further in Chapter 17.

WAVEGUIDES   It seems likely that the next major step forward in expanding the arteries that carry large numbers of voice or other channels across the country may be the use of waveguides.

A waveguide is, in essence, a metal tube down which radio waves of very high frequency travel. There are two main types of waveguide, rectangular and circular. Rectangular waveguides have been in use for some time as the feed between microwave antennas and their associated electronic equipment. It is normal to see a waveguide going up a microwave tower to the back of the dish that transmits the signal. They are not used for long-distance communication and are rarely employed for distances over a few thousand feet. They consist of a rectangular copper or brass tube, 15 in. across or smaller (Fig. 14.3). Radiation at microwave frequencies passes down this tube.

Circular waveguides are pipes about 2 in. in diameter. They are constructed with precision and are capable of transmitting frequencies much higher than rectangular waveguides, or the other media discussed. Figure 14.4 shows the construction of a Bell System waveguide [2], which is

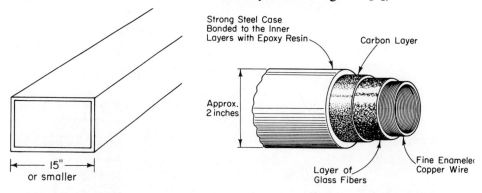

Figure 14.3.   Rectangular waveguide.

Figure 14.4.   Helical waveguide.

referred to as a helical waveguide because a fine enameled copper wire is wound tightly around the inside in a helix. The wire is surrounded by a layer of thin glass fibers and then by a carbon layer. The whole is encased in a strong steel case and bonded to it with epoxy resin. Figure 14.5 shows a photograph of this pipe. The purpose of this construction is to attenuate undesired modes of wave propagation. Waveguides cannot have sharp bends in them; however, they can have large radius bends like a railroad track [3]. These gentle bends cause little loss of signal strength. Sharp corners need a junction between two waveguides with a repeater.

The frequencies used on wire pairs, coaxial cables, and waveguides, and the different levels of attenuation associated with them, are shown in Fig. 14.6. It will be seen that coaxial cables permit transmission at a much higher frequency than wire pairs, and waveguides are much higher than coaxial cables. The loss in waveguides built so far actually becomes less as frequency increases up to frequencies of about 100,000 MHz, as shown. Theoretically, it should continue to lessen indefinitely as frequency increases, although an upper limit is set by today's engineering.

Helical waveguides have been operating in experimental circuits for some time. The noise level encountered on the Bell System waveguide now operating is low enough for circuits of several thousand miles to be constructed; for example, coast-to-coast links in the United States. The first waveguides in public use are to be installed shortly on a test basis; they can carry 230,000 voice channels. It seems probable that future waveguides will employ digital transmission (pulse code modulation) entirely.

Figure 14.5. AT & T advertisement showing their helical waveguide.

A two-inch empty pipe
can carry 230,000 telephone conversations.

The pipe is no bigger than your wrist.

Yet what really makes it news is that there's absolutely nothing inside.

Except room for 230,000 simultaneous telephone conversations.

In the years to come, millimeter waveguide pipe will be buried four feet underground. In a larger cradling pipe to give it protection and support.

It'll also have its own amplifying system about every 20 miles. So your voice will stay loud and clear.

Even after 3,000 miles.

Yet this little pipe is capable of carrying a lot more than just conversations.

It can also carry TV shows. Picturephone® pictures. Electrocardiograms. And data between thousands of computers.

All at once.

The American Telephone and Telegraph Company and your local Bell Company are always looking for new ways to improve your telephone service.

Sometimes that means developing a better way to use two inches of empty space.

Figure 14.5. Cont.

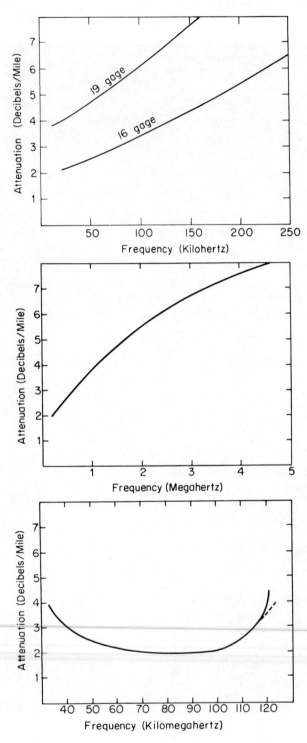

Figure 14.6. Attenuation on cable and waveguides.

**RELATIVE
COSTS**

Figure 14.7 shows the relative costs of twisted-wire-pair cables, coaxial cables, microwave radio relays, and helical waveguides [1]. The costs used in plotting this chart were based on today's technology. Satellite transmission is going to override some of these costs but cannot be plotted on the same chart because its costs are not proportional to distance, as are land circuits. This subject will be discussed in the next chapter.

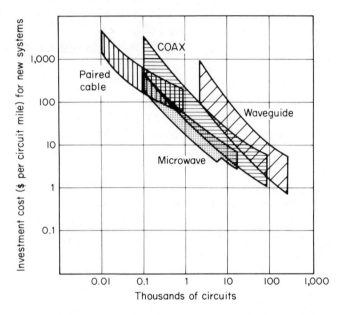

Figure 14.7. Cost trends in terrestrial transmission (1970). *Redrawn from Reference 1.*

Wire-pair cable will probably be the dominant mode of telephone transmission in the next decade for routes that carry less than 500 voice circuits (or other signals equivalent to this total bandwidth), provided they are not too long, for wire pairs have range limitations. The greatest use of wire-pair cable will be over distances less than 10 miles.

Microwave radio giving point-to-point line-of-sight relay of signals becomes the cheapest mode for between about 500 and 15,000 circuits. Since Microwave Communications, Inc. obtained a long-fought-for ruling from the FCC allowing them to establish a microwave link from St. Louis to Chicago and market its bandwidth, there have been many proposals for new types of common carriers offering the use of microwave, including the Datran proposal described earlier. Over mountainous terrain, as well as lakes and rivers, microwave circuits are easier to construct than cable circuits. In cities and on approaches to cities, however, spectrum congestion may become a problem.

Coaxial cable gives the lowest cost for between 15,000 and 80,000 circuits. There will be many links of these volumes in the years ahead. Voice transmission is expected to triple by 1980, and to this will be added the likelihood of Picturephone circuits requiring more total bandwidth than the total for voice, an increase in television transmission, and a great increase in data transmission. AT & T has predicted that by 1980, 90 % of its long-haul circuitry will be coaxial cable.

By 1980 there will be many routes with traffic greater than 80,000 telephone conversations or equivalent. The helical waveguide may dominate on these high-capacity links.

LASERS   Still higher in the electromagnetic spectrum than the frequencies of the foregoing transmission media (Fig. 13.1) is the laser, operating at the frequencies around those of light, and for experimental earthbound communications having an information rate that is low in comparison with their theoretical capacity. It seems probable, however, that eventually they will provide the means of building channels of enormous capacity—immensely greater even than the helical waveguides. It has been said that lasers portend a revolution in telecommunications as fundamental as the invention of radio. Lasers have been used successfully for transmission from space vehicles.

Laser stands for "Light Amplification by Stimulated Emission of Radiation" and was preceded by *maser,* meaning "Microwave Amplification by Stimulated Emission of Radiation." A laser produces a narrow beam of light that is sharply monochromatic (occupies a single color or frequency) and coherent (all of the waves travel in unison like the waves traveling away from a stone dropped in a pond). Normal light, even that of one color, consists of a small spread of frequencies and waves that are incoherent, bearing a random position relative to one another. A lasing action often produces more than one "spectral line," and in that case each one of them is sharply monochromatic.

An analogy with sound waves is somewhat inexact, but it will help the reader to visualize the difference between a laser beam and an ordinary light beam. The sound from a tuning fork consists of waves that are of one frequency and that are reasonably coherent. On the other hand, if I put a hammer through my apartment window, the sound waves would be neither monochromatic nor coherent. The former may be compared with a laser beam, the latter with ordinary light. The laser or maser beam is formed by a molecular process somewhat analogous to the tuning fork. It is possible to make certain molecules oscillate with a fixed frequency in much the same manner as the tuning fork.

The electrons in an atom can only move in certain fixed orbits. Associated with each orbit is a particular energy level. The electrons can sometimes be induced to change orbits, and when this step happens, the total energy associated with the atom changes. The atom can therefore take on a number of discrete energy levels—a fact that is well known today from quantum mechanics. Certain processes can induce the electron to jump from one orbit to another or, to state it another way, to induce the atom to switch from one energy level to another. When this happens, the atom either absorbs or emits a quantum of energy. In this way, light, radio waves, or other electromagnetic radiation is emitted in discrete quantums.

When ordinary light is emitted, the mass of molecules switch their energy levels at random. A random jumble of noncoherent waves is produced. Under the lasing action, however, the molecules are induced to emit in unison, the substance oscillating at a given frequency, and a stream of coherent waves at this single frequency results. This could be either a microwave frequency (maser) or a light frequency (laser).

When a laser beam produced by certain lasing molecules falls on other molecules of that type, it can induce oscillation in them. A form of resonance is set up. The reader might imagine a huge pendulum much too heavy to move far by a single hard push. If he gives a series of relatively gentle pushes, however, he can set it swinging. He may go on pushing at just the right point in the swing, and the length of the swing increases until the pendulum builds up great power. This is resonance. His gentle pushes have been amplified into massive oscillations. In a similar manner (and again the analogy is helpful but not exact), a weak laser beam can fall on a lasing substance and cause resonance in it. It sets the molecules oscillating so that a powerful laser beam is emitted. The laser beam has thus been amplified. In this way, a very intense beam of a single frequency or several single frequencies can be emitted.

A beam of ordinary light, even a beam that we describe as monochromatic, actually consists of a small spread of frequencies, each of which would be bent slightly differently by a prism or lens. A laser beam, however, is not dispersed by a prism and optical arrangements can be built for it so precisely that a beam of laser light can be shone onto the moon and illuminate only a small portion of its surface. A beam can be concentrated with a lens into a minute area, and the intense concentration of energy into such a small area causes very localized heating to occur. A cutting or welding tool is provided with a miniature precision far beyond the dreams of Swiss watchmakers. The surgeon has a microscopic scalpel; the general a potential death ray.

To summarize, laser light differs from ordinary light in four characteristics. The wide variety of applications of the laser are based on one or more of these four factors.

1. Laser light is sharply monochromatic. It may, however, have more than one monochromatic frequency (color). By using filters, a single frequency can be separated from the others.
2. The light is coherent, the waves being regularly arranged and in phase with each other.
3. It can have very great intensity, so great that it could blind a person instantly.
4. It has very low dispersion. A beam shone onto the moon illuminated an area only 2 square miles across. An attempt is being made to send a laser beam to Mars, one hundred million miles away. On earth, lenses can focus the laser into *very* tiny intense spots.

For telecommunication use, we have a beam of great intensity that is highly controllable and that can be amplified. Whereas today's microwave beams disperse over an angle of about 1 degree, a laser beam could be almost exactly parallel. However, the most exciting fact about it is that its frequency is about 100,000 times higher than today's microwave and its potential bandwidth many hundreds of times greater. If we can learn how to superimpose the information on the beam (modulate the beam) at a sufficiently high rate, laser communication links in the future may well carry many hundred times as much information as today's microwave links.

The biggest problem is how to modulate the beam. Its intensity or some other property must be varied sufficiently fast to convey the large quantity of information. A large number of different lasing actions could be used, and a variety of different techniques for making the laser beam carry information has been suggested. It is not certain yet which of these will eventually prove the best for telecommunications. To give a crude example of what is meant by modulation, the beam could be varied by means of a series of rapidly moving shutters, diffraction gratings, or mirrors. The variation would be used to convey information. This mechanical process, however, would be extremely slow, compared with the capacity of the beam. The same function must be performed electronically at enormous speed. One possible way of doing so would be to use electric field absorption. The absorption region of a semiconductor can be transferred to a different wavelength when an electric field is applied. Varying the field can thus be used to vary the absorption of a laser beam, and so its amplitude may be varied at rates as high as several billion (10^9) times per second.

Another possibility is to use a phenomenon called *Pockel's effect*. Here the beam is shone through a transparent piezoelectric crystal. When an electric field is applied, it strains the crystal and rotates the plane of polarization of the laser beam. Again the beam has been modulated at rates of several billion cycles per second.

A laser can be made to produce a series of very narrow pulses of light, and this process is the basis of a transmission system being developed at the Bell Laboratories. When a laser emits different frequencies, each monochromatic and coherent in phase, these frequencies interfere with one another, in some places cancelling each other out and in certain places reinforcing each other to form a narrow high-intensity pulse. The larger the number of frequencies used, the narrower the pulses can be. Exceedingly narrow pulses can be produced, spaced relatively far apart in time. These pulse trains can be made to carry digital signals and hence can be used for pulse code modulation. Because the pulses are relatively far apart, there is time, using present-day techniques, to modulate them; and because the pulses are very narrow, several pulse trains can be interleaved (time-division multiplexed).

The first experimental system at the Bell Laboratories used a gaseous helium-neon laser that produced a pulse stream of 224 million pulses per second. A laboratory terminal was built capable of interleaving and transmitting four such pulse streams, giving a total of 896 million bits per second. The pulses are narrow enough, however, to interleave 24 such pulse streams. Furthermore, two light beams polarized at right angles to one another can be transmitted at once, giving 48×224 million $= 5.376$ billion bits per second.

Figure 14.8 shows a terminal designed to demultiplex a 5.376-billion-bits-per-second pulse stream. The multiplexing process is the converse of what happens in the figure. A polarizing filter at the top of the figure splits the stream into two 2688-million-bits-per-second streams, which then enter two identical demultiplexing units. The demultiplexing units consist

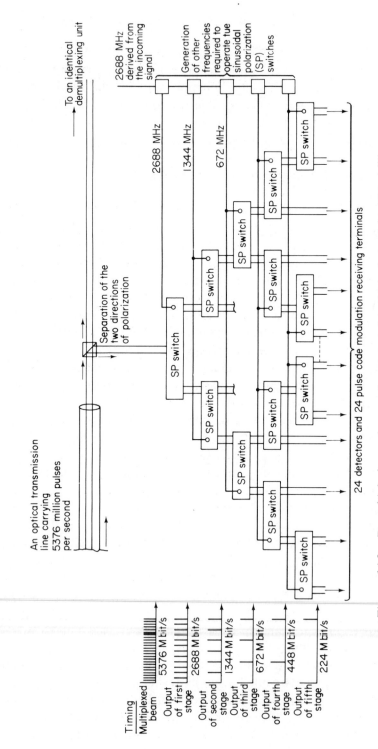

Figure 14.8. Demultiplexing a laser pulse stream in which forty-eight 224 million bit/second signals are sent as pulses of the same frequency.

216

of a cascade of polarization switches. The switches consist of crystals that rotate the plane of polarization of light, depending on the voltage applied to them. They are used for selectively removing the pulses of unwanted channels and are operated by a sinusoidal voltage. There must be very exact synchronization and so timing information is derived from the input pulse stream.

Other lasers today can give much higher pulse rates than the above gas laser. Solid state lasers, in general, can give much narrower pulses and there are many types of solid state lasers. A promising one uses yttrium aluminum garnet (YAG) doped with neodymium and generates pulses of 35 picoseconds (35×10^{-12} seconds) duration. This laser may be used to give multiplexed PCM channels at rates as high as 1.5×10^{10} bits per second.

The shortest experimental laser pulse that has been produced at the time of writing is 0.4 picosecond in duration. Theoretically, it could give a pulse stream of more than 10^{12} bits per second.

Many new developments are probable in this research. The bandwidth theoretically available could carry many more, even, than 10^{12} bits per second, but much work needs to be done on developing the new types of optical circuitry.

LASER PIPES   Although the laser beam is sometimes visualized as being shone through the air as a pencil-thin beam of light, as with the optical systems we discussed in the previous chapter, such a scheme would have many disadvantages. For example, it could be interfered with by anything that would disturb ordinary light—fog, snow, or boys with kites. A bird flying through the beam would obliterate 10 million bits. A grave problem in making such a system function correctly would be the variations in the atmospheric path used. Instead, the beam is likely to be sent through a pipe containing a bundle of optically transparent fibers. If a fiber is surrounded by a substance of lower-refraction index, such as air, light passing down it will be totally reflected by the edges of the fiber. This total internal reflection occurs in a similar manner at the surface of a pond. If you put your head under the water and look at the surface some distance away, it will appear to be a totally reflecting mirror. The ray of light is not refracted out of the pond at all because of its low striking angle but is totally reflected back into the water. The laser beam travels down a transparent fiber and is confined within the fiber by total internal reflection. It will be absorbed somewhat by the fiber, and it will have to be periodically amplified. Lasers of many different frequencies could travel together down the same fiber, and a bundle of such fibers could occupy one pipe.

Flexible optical fiber cables can be constructed. Figure 14.9 shows the

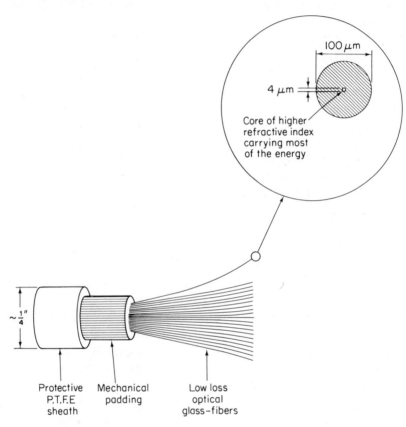

Figure 14.9. Possible form for a flexible optical cable. *Re-drawn from Research, a publication of the British Post Office.*

possible form that one might take. It might contain many thin glass fibers, each with a center core only a few micrometers in diameter, surrounded by glass of slightly lower refractive index, about 100 micrometers in diameter. Coherent light would be propagated as a surface wave on the inner core. By making the glass to exceptionally high standards of purity it might be possible to reduce the loss in such a fiber to 30 dB/mile so that digital repeaters could be approximately two miles apart. Glass of this quality can be made but drawing it into suitable fibers presents considerable difficulty. If such problems are solved readers of this book may live to see optical cables going into their homes, perhaps replacing their previous coaxial CATV cable.

A cross-country laser pipe of very high capacity would benefit from being rigidly constructed and may be built in straight-line segments. The civil engineering task of building the future laser highways will not be trivial. It is possible that they will use the same paths as those built earlier for helical waveguide channels.

It has been said that such a laser system has the potential ability to carry all the information carried by all the telephone lines in the world today at the same time.

REFERENCES

1. Eugene V. Rostow, "A Survey of Telecommunications Technology," the President's Task Force on Communications Policy, Staff Paper No. 1, Washington, D.C., 1969.
2. A. P. King, and G. D. Mandeville, "The Observed 33 to 90 kmc Attenuation of 2-Inch Improved Waveguide," *Bell System Tech. J.*, September 1961.
3. D. T. Young, "Measured TE_{01} Attenuation in Helix Waveguide with Controlled Straightness Deviations," *Bell System Tech. J.*, February 1965.
4. H. H. Steier, "Attenuation of the Holmdel Helix Waveguide in the 100–125 kmc Band," *Bell System Tech. J.,* May 1965.
5. Tracy S. Kinsel and Richard T. Denton "Terminals for a High-Speed Optical Pulse Code Modulation Communication System: II Optical Multiplexing and Demultiplexing," *Proceedings of the IEEE*, Vol. 56, No. 2, February 1968.
6. R. T. Denton, "The Laser and PCM," *Bell Laboratories Record*, June 1968.

H.M.S. Agamemnon laying the first transatlantic telegraph cable in 1858. The cable had a spectacular but short life. After some of the toughest financial wrangling of the nineteenth century and more than a year of heartbreaking failures on the high seas, it was eventually laid successfully. The signals that trickled through it were so minute that only the most sensitive suspended-mirror galvanometer could detect them. The first message of ninety words from Queen Victoria took 16½ hours to transmit. Press headlines were sensational beyond precedent. However, some days after the Queen's message, the insulation of the cable failed and it never worked again. It was eight years before another cable was laid. One American newspaper called the cable a hoax, and an English writer "proved" that it had never been laid at all.

15 COMMUNICATION SATELLITES

The Satellite Telecommunications Subdivision of Electronic Industries Association (EIA) hereby calls for the aggressive pursuit of a domestic telecommunications satellite system with implementation to be effected at the earliest possible date. We urge that all appropriate responsible governmental, legislative, and industrial activities vigorously work toward this end without further delay. Problems of ownership, financing, frequency allocation, etc., must be resolved in a positive manner, but without unduly compromising the United States' worldwide leadership position. It is in the national interest to resolve the terrestrial satellite interface and interference problems as rapidly as possible by an early implementation of a U.S. domestic communications satellite project. The recently approved Canadian and Soviet National Satellite Telecommunications Systems are positive proof that other communities are proceeding to assume a lead role in the implementation of national satellite telecommunications. The Satellite Telecommunications Subdivision of EIA feels that the United States can ill afford to be preempted—[1]*Resolution of the EIA Satellite Telecommunications Subdivision, April 25th, 1968.*

Many times in this century technology has raced ahead of informed forecasts made for it. Forecasts made in 1960 looking a decade ahead failed to predict the rate of growth data transmission, real-time systems, time sharing, computer capabilities, and, in particular, they underestimated the success of the communications satellite.

The inability to foresee developments and uses of technology sometimes results in inappropriate laws being passed. The United States Communication Satellite Act of 1962 was one such law because when this was passed satellites were seen as being of value for international transmission only.

A communication satellite provides a form of microwave relay. It is high in the sky and therefore can relay signals over long distances that would not be possible in a single link on earth, because of the curvature of the earth, mountains, and atmospheric conditions. The first communication satellites were in relatively low-altitude orbits and consequently were overhead for only a brief period. In 1965 the Early Bird satellite proved the viability of using satellites in orbits sufficiently high that they revolve around the earth's axis in exactly 24 hours. These are referred to as *synchronous* satellites. If launched in a synchronous orbit over the Equator, the satellite appears to hang stationary above one point on earth, 22,300 miles away.

Ten years ago it seemed doubtful whether rocketry would achieve such accurate positioning in the near future or whether electronics could achieve a sufficiently noise-free link via a small solar-battery-powered object this far away. Orbital locations were regarded as a hostile environment for electronic circuitry. Today they seem more benign. The satellite is in very high vacuum, far from man-made noise sources, and is at a near-constant temperature. The wear and tear caused by expansion and contraction, and by atmospheric corrosion, are largely absent. Taking advantage of the environment, we will produce long-lasting and efficient satellites. In the decades ahead, a major industry will evolve concerned with the positioning, operation, and maintenance of large objects in synchronous earth orbit.

Satellites have long appeared attractive for reaching parts of the earth that are inaccessible by terrestrial links. A ground station in Africa or South America can be erected quickly at moderate cost, and these locations can be linked to the world's telecommunications networks. At first, such usage was foreseen as the main use for satellites. In the 1960's, however, it became clear that satellites could provide a lower cost means of sending telephone calls or television across the oceans than the expensive submarine cables. Such links proved very satisfactory in operation.

Now, with the economics of satellites improving still further, they are clearly going to form an important part of the transmission networks on land. Within the United States, for example—if engineering rather than politics is permitted to prevail—they will form an important part of the long-haul telephone network, will provide data transmission facilities, and will be extensively used in television distribution.

To travel 22,300 miles into space and back takes the signal a quarter of a second. Half a second's delay occurs, therefore, before you hear a reply in a telephone conversation if a satellite is being used for transmission in both directions. AT & T (which does not own any satellites) has sometimes advocated that this delay is too long and is inacceptable to telephone users. On many of today's satellite telephone calls, the satellite is used for one direction of transmission and subocean cable is used for the other direction. The result is a total delay slightly greater than a quarter of a second. It also sometimes results in the disturbing effect of person A being able to hear person B very distinctly, as though he were in the next room, but person B being able to hear person A only poorly, and hence feeling that there is a need to shout. The more the clearly heard person shouts, the more the poorly heard person reacts by talking softly, and vice versa.

Figure 15.1 shows the results of a set of experiments in which delay was added artificially to telephone calls and the callers were asked to comment on the quality of the lines. It will be seen that slightly more than 10 percent complained even when there was no delay. This percentage increased only slightly when a delay was introduced equivalent to a single satellite round trip (up and down). When two round trips were included the number of complaints had risen to about 20 percent; 80 percent did not complain. When the delay reached 750 milliseconds (3 round trips), the complaints were increasing rapidly. The actual delays on a satellite call are slightly longer than those shown because the propagation time on the interconnecting land links have to be added. However, the figure suggests that the public could become accustomed to telephone calls that travel over two satellite hops, though not three. Where satellites are used in the United States, the switching will have to be organized so as to prevent interconnection over more than two satellite links for the round trip.

The delays cause problems mainly when the speakers are responding to one another very rapidly, or when they both attempt to speak at once. If they both appreciate that the response delay will occur, they will probably learn to exercise slightly more discipline in their interaction.

THE EFFECT OF SATELLITE SIZE   Two factors affect the performance of synchronous satellites: their size, which determines their power among other things, and the frequencies they use, which determine their bandwidth or signal-carrying capacity and the narrowness of the radiated beams, which in turn determines the maximum number of satellites that can be placed in orbit.

In general, the larger the size or power of the satellite, the smaller the earth antenna needed. The first small satellites need very large antennas, as shown in Fig. 15.2, and their earth stations were very expensive. As the satellite sizes increase, the earth-station size and cost will drop until the ultimate situation is reached when a house or even a television set is its own earth station and has a small dish-shaped antenna pointed at the satellite. There are a number of possibilities in between those two extremes.

The main types of synchronous communication satellites can be divided into five groups:

1. *The point-to-point relay satellite*

All the nonmilitary satellites in operation at the time of writing are of this type. They transmit between large fixed earth stations, such as that

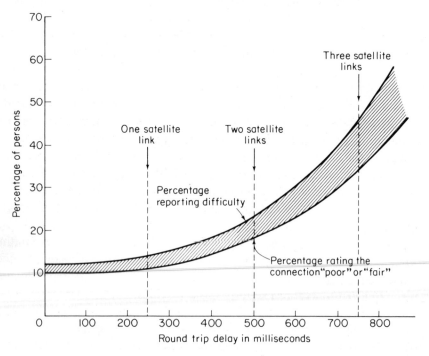

Figure 15.1. A subjective assessment of the effect of transmission delays on telephone callers. The public would probably become used to satellite circuits with two round trips.

Figure 15.2. A transportable satellite earth station. It can be driven to a suitable location in the truck at the right.

in Fig. 15.1. High-capacity groups of frequency-division-multiplexed signals are relayed between these points, as over the common carriers' major terrestrial channels. The satellites provide two-way links and can carry voice, television, data, or any other signals that can fit into the multiplexed bandwidth, as on earth. The link can thus be regarded as part of the common carriers' routing facilities.

The early satellites have been small in size and power; hence the earth stations have been expensive, typically $4 to $8 million. This factor makes the link competitive only over very long distances on land, although it is lower in cost across the oceans than today's submarine cables. As satellite power increases, the earth-station cost will drop, and during the 1970s it will drop to less than $500,000. In addition, as we shall see later, the capacity of the satellites can be increased. These two factors will make them competitive with land lines over shorter distances. The economics indicate that they should eventually become a major part of national systems, such as the Bell System and the Datran System.

The two main factors that make the earth station expensive today are the large size of the antenna (typically about 100 ft in diameter), the fact that it must be steerable, and the steps needed to minimize the transmission noise. The thermal noise generated in transmission is proportional to the absolute temperature, and so the circuitry in the transmitting station is cooled by liquid helium. With future satellites of not much higher cost, the same bandwidth will be relayed by earth stations with smaller (e.g., 30 ft), nontracking antennas and uncooled receiver amplifiers. Figure 15.3 shows what a proposed set of point-to-point linkages would do for South America.

2. Multipoint relay satellites

If satellites are to become a major part of common-carrier networks, it

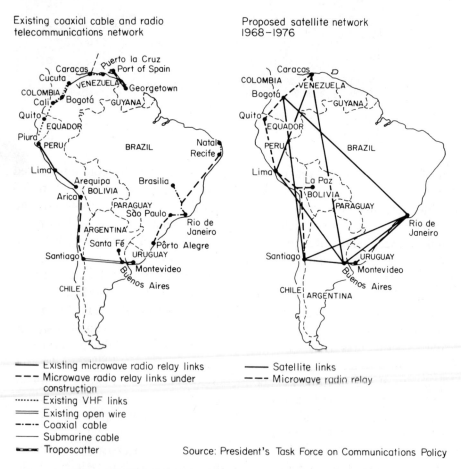

Existing coaxial cable and radio
telecommunications network

Proposed satellite network
1968–1976

——— Existing microwave radio relay links
– – – Microwave radio relay links under
      construction
········ Existing VHF links
═══ Existing open wire
–··–·· Coaxial cable
——— Submarine cable
▬▬▬ Troposcatter

——— Satellite links
– – – Microwave radio relay

Source: President's Task Force on Communications Policy

Figure 15.3. What point-to-point satellites can do for South America.

would be advantageous for many earth stations to be able to use them, with calls being switched between different stations and different satellites, as desired.

In such a system, the satellite will relay many different groups of signals. Each earth station will transmit only one or a small number of the groups. It may receive all the transmitted groups and then send only those addressed to it over the terrestrial trunks.

A system of this nature would be more efficient if highly directional antenna beams could be used between satellite and earth station, so that the same frequency could be reused by geographically separate earth stations. Distant satellites would also reuse the same frequencies. Today's satellites operate at microwave frequencies from 4 to 6 GHz. At these frequencies, a highly directed beam can only be achieved with a very large antenna. As we shall discuss later, satellites of the future may use millimeter waves with which highly directed beams can be sent. Intelsat IV will employ two such directive microwave antennas, although frequencies will not be reused initially.

Some multipoint relay systems of the future will probably serve a large number of the earth stations. With a large satellite, the earth stations may be very small and inexpensive, without a 15-ft antenna, say, and uncooled electronics. They may be mobile stations where circumstances warrant mobility—for example, in underdeveloped countries, on ships, in the Arctic, and in military situations. Small earth stations may have only a small number of voice channels. Stations may operate, indeed, with only one voice channel. The bandwidth will be used much less efficiently if small groups of channels or single channels are transmitted. It is like comparing the bandwidth per voice channel needed on mobile radio (25 to 100 kHz) with that on common-carrier lines (4 kHz). Voice-channel spacing of 100 kHz is used on some present-day satellite designs, as in broadcasting. This situation is clearly much more wasteful of bandwidth than if trunk groups are relayed, within which the channel spacing is 4 kHz.

Schemes have been proposed for allocating the satellite capacity on a demand rather than a fixed basis, which would be of value where the demand is not continuous. Such allocation may permit the linking up of many locations in developing countries or isolated parts of the world with little traffic. Figure 15.4 illustrates such a scheme with three satellites [1]. The satellites in this proposal have a bandwidth of 500 MHz. All the channels carry one two-way voice signal and need an assignment of 100 kHz. The system thus has a capacity of 5000 voice channels—much

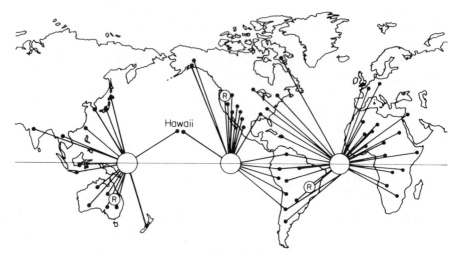

Hawaii

Legend : • Demand access earth terminal, ⓡ Routing center

Figure 15.4. Three multipoint satellites giving world coverage to lower-volume telephone users on a demand-assigned basis. The three routing centers have digital links to the satellites for controlling the assignment of the channels.

less than for a common-carrier, high-density relay satellite. The control of the demand assignment is done by a routing center. Each satellite has one such center, connected to it by a data link, as shown.

3. A distribution satellite

This is a satellite for distributing television, radio, or other one-way signals, such as news services or stock market information. The signals will be picked up by relatively inexpensive antennas, which relay them to local broadcasting stations. One such system proposed by Comsat for television distribution in the United States uses antennas that are 30 to 40 ft high and that cost about $100,000.

Figure 15.5 shows a possible configuration [1]. There are a relatively small number of transmitting stations, two national and the others regional. The number of receiving stations is much larger, and these stations are substantially lower in cost than the transmitting stations. The beamwidth of the satellite is 3 degrees, which gives suitably high transmission power. Four different antennas on the satellite are used to cover the four time zones of the continental United States. (The continental United States subtends 7.5 degrees at the satellite.)

4. *Direct broadcasting satellite*

Increasing the power of the satellite can permit direct television pickup on home antennas. The antenna for this operation may be about 5 ft in diameter. A still more powerful satellite could broadcast direct into a rabbit-ear antenna on the living room set, but a more probable arrangement is rooftop antennas, as shown in Fig. 15.6. Such antennas have been estimated to cost from $50 to $150 each.

Figure 15.7 gives the estimated cost relationships in direct satellite broadcasting, using present-day technology [2]. These figures were calculated assuming a system designed for high-quality reception and using the 800-MHz UHF frequency band. The five-year capital costs per household of the satellite and earth stations were as follows [2]:

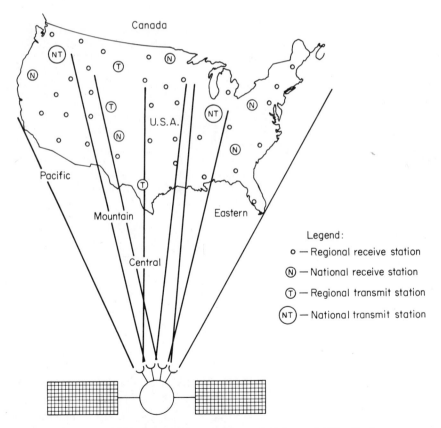

Figure 15.5. Domestic one-way multichannel TV distribution system showing national and regional stations.

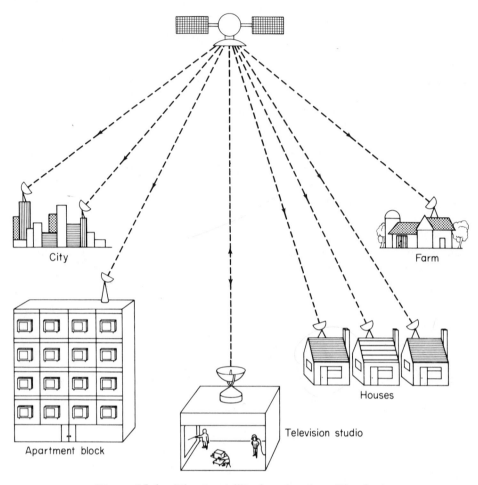

Figure 15.6. Direct satellite broadcasting. Five-foot antennas on the roof-tops receive television or other one-way signals directly from a large relay satellite.

Number of Viewers	1 Channel	12 Channels	24 Channels
1 million	$60	$225	$315
10 million	36	93	113
50 million	27	43	63

5. A satellite for cable TV

A satellite serving a small community will probably be preferred to one serving individual homes. This satellite would need less power and its receiving station might cost, say, $5000 to $10,000. The antenna might often be at the head of a cable TV system. In this way, the large numbers of cable channels discussed in Chapter 5 may be filled, sometimes, one hopes, with educational television, sometimes with foreign television. Where cable TV is not yet installed, a local TV transmitter may relay the signals.

It is by this means that television will first reach the villages of poorer countries. Both Russia and the United States are planning direct broadcasts, and their impact could be immense. NASA, for example, is to broadcast educational programs to India for 4 to 6 hours a day, using their Applications Technology Satellite, ATS-F. In this experiment, India will initially have 5000 village receivers, manufactured in India.

COSTS OF
SATELLITE
LINKS

The costs of satellite circuits with one continent are essentially independent of distance, whereas those of terrestrial links are proportional to distance. A satellite system thus has an economic advantage on long-distance routes. It also has an advantage on routes with high volume because the earth-station cost is spread over more traffic.

As the satellite capacities become bigger, as the traffic volume rises, and as the earth-station cost drops because of greater satellite power, so

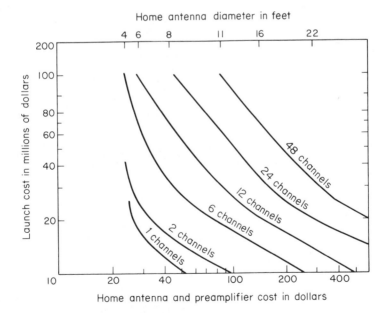

Figure 15.7. Cost relationships in direct satellite broadcasting using today's technology. *Source: President's Task Force on Communications Policy (2).*

the breakeven point between satellite links and land links occurs at shorter distances. Figure 15.8 gives some typical costs of satellite links, using today's technology [3]. The figure relates to satellites of three different sizes, having capacities of 6750, 13,500, and 40,500 circuits respectively.

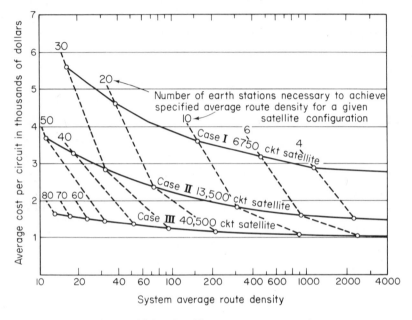

Figure 15.8. Satellite systems cost trends.

It will be seen that costs drop substantially as satellite capacity increases. The largest satellite of the three has a lower cost per circuit when the average route density is 24 telephone channels than the second satellite has for a density of 600. This dramatic drop differs from terrestrial economics, in which route density is the primary cost determinant.

As previously mentioned, the average cost of adding a circuit mile to the Bell System long-haul lines is falling rapidly and may drop to $1.40 by the end of the 1970s. Satellite cost will probably drop at a faster rate. If we take $1400 as being a typical circuit cost for the larger satellite in Fig. 15.8, it will be seen that a domestic common-carrier satellite system already appears attractive economically for longer haul lines.

Several other factors in satellite economics should be considered. First, the life of the satellite will not be as great as that of land lines. It may be as low as 5 years, and a new launch will then be required. On the other hand, routing via the satellite can be varied, whereas on land lines the capacity of routes is fixed. Thus the satellite will normally have a higher utilization than terrestrial trunks. Again, satellites are particularly economical for the wide distribution of one-way signals, such as television, because the receiving earth stations can be low in cost.

The variable routing may be very valuable for some services. For example, video conferences require very high capacities from certain locations for the duration of a conference only. It is possible that major cities may have specially equipped conference halls linked by satellite. Computer-assisted instruction networks may also be highly variable in demand for channel capacity.

MICROWAVE RADIO INTERFERENCE Satellites operating today use microwave frequencies and problems can arise with interference between them and terrestrial microwave systems. The proposed Canadian satellite system uses a frequency of 6 AHz to transmit to the satellite and 4 AHz to transmit from the satellite to earth. These are frequencies used by terrestrial microwave links and interference might occur as shown in Fig. 15.9. The degree of interference depends upon the relative positioning of the interfering antennas. If they are far apart and the antenna

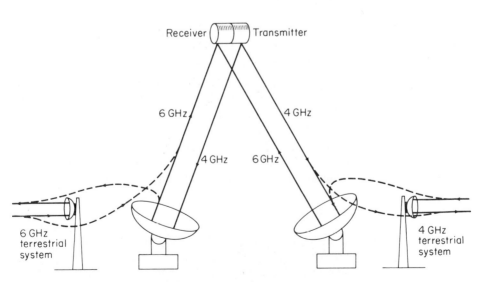

Figure 15.9. Possible interference between satellite and terrestrial links.

dishes do not point at one another there will be no problem. It is expected that interference will be avoided on the Canadian system by placing the satellite earth stations 50 to 100 miles from urban centers that are linked with terrestrial microwave chains. In the United States there would be more of a problem because there are more microwave systems. Careful positioning of antennas and design of directional antennas is necessary. It seems likely that certain frequencies will be allocated exclusively to satellite use.

At higher frequencies, especially those above 10 AHz, radio interference is not likely to be a problem. UHF frequencies have been discussed for satellite use, but the UHF spectrum is a resource in short supply that will be increasingly needed for terrestrial applications. Interference between terrestrial and satellite transmissions using the same frequency would be more of a problem in the UHF band than in the microwave band.

THE FREQUENCIES EMPLOYED

The 1970's may bring a move into millimeter-wave frequencies for satellite transmission. The big disadvantage of millimeter-wave frequencies (above 10 GHz) is that they are heavily absorbed by cloud, rain, and to some extent by the atmosphere itself. Because of the absorption, higher satellite power will be needed.

Nevertheless, the advantages of using them instead of microwave are great:

1. The bandwidths available are much higher. Satellites of very high channel capacity could be constructed.

2. Microwave interference increases as the radiated power from the satellite goes up, as it must do if we are to achieve low-cost earth stations. The microwave band, in other words, will almost certainly become congested, as are the VHF and UHF bands today.

3. The beam from the satellite can be much more narrowly directed, which means that the received signal will be stronger. It also means that the same frequencies can be reused many times at different earth stations.

4. The number of satellites in orbit can be much greater. With satellites using the same microwave frequencies, it has been estimated that not more than a total of 75 could be in orbit without causing undue frequency interference (Fig. 15.10). The spacing can be closer using millimeter waves. It has been estimated that 95 satellites spaced 1 degree apart could serve the North American continent via the same

frequency band (Fig. 15.11). Other satellites using other frequency bands could also be in orbit, and many such millimeter-wave families of satellites may eventually hover over the equator.

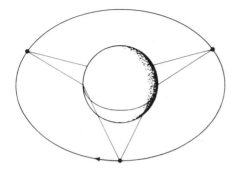

(a) The minimum number of satellites that will give world coverage (not counting the poles) is 3

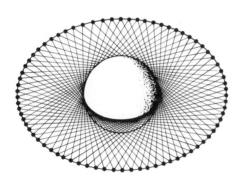

(b) The maximum number of satellites operating at today's frequencies is about 75. More than this will lead to undue frequency interference. However many more satellites could be used at millimeter wave frequencies

Figure 15.10

Another possible form of transmission, much higher in the electromagnetic spectrum, is the laser. The main problem with the high frequencies at which lasers operate is again the fact that absorption takes place in clouds and heavy rain—much worse absorption than occurs with millimeter waves. Light, as humans can observe, is blotted out by clouds. So are the other frequencies that are above or below the frequency of light. There are, however, "windows" in this part of the spectrum, at which the absorption is less. One such window occurs at the frequency of the CO^2 laser—30,000 GHz (light frequencies are around 300,000 GHz). This window is 40 GHz wide, and the attenuation caused by clouds and atmosphere is less than at any frequency in the visible or ultraviolet ranges. These frequencies might be used in the future for satellites. They offer a high bandwidth and no interference problem. The CO_2 laser may be used for satellite transmission.

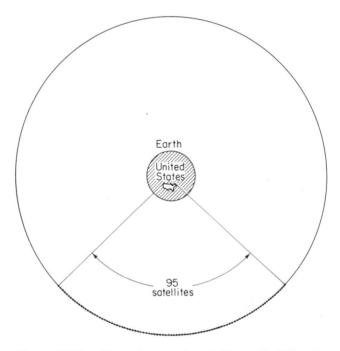

Figure 15.11. Using frequencies of 10 to 40 GHz, the maximum number of "stationary" satellites that could serve the United States, operating on the same frequency band is about 95. This is much higher than with microwave frequencies (Fig. 15.10). Still more satellites could operate alongside these at other frequencies. This diagram is drawn approximately to scale.

ABSORPTION OF MILLIMETER WAVES To overcome the disadvantage of millimeter waves, the absorption caused by rain, cloud, and, to a lesser extent, by the air and water vapor, higher-transmitter powers will have to be used. Even then the beam will occasionally be attenuated badly or even blotted out by intense storms. Higher-transmitter power is also needed because the noise temperature increases with the increase of frequency.

Figure 15.12 shows the effect of rainfall on satellite transmission, plus the fact that the attenuation increases rapidly with frequency for millimeter waves. The vertical axis shows the attenuation that will be exceeded 0.01% of the time (slightly less than one hour per year). These figures were estimated for a site in the United States having a rainfall of 74 cm/year [1].

The attenuation due to rain is less if the transmission path to the satellite is nearly vertical, for a shorter distance is traversed in the storm. Figure 15.12 also shows the effect of varying the angle of elevation. For domestic transmission, a high angle of elevation will be possible. Also, most rain occurs below an altitude of 2 km, so there will be a great advantage in having an earth station on a mountaintop.

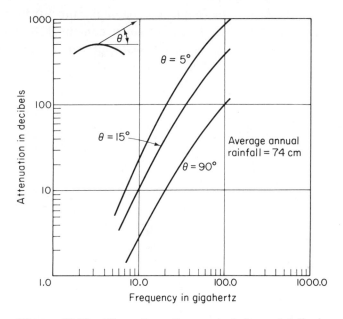

Figure 15.12.   The attenuation caused by rainfall, in satellite transmission at different frequencies. These curves are estimates for a site in the United States with an annual rainfall of 74 inches and show the attenuation figure that would be exceeded 0.01 per cent of the time. This is shown for three different elevation angles. Rainfall attenuation severely affects millimeter wave transmission. *Reproduced from reference 1.*

Gentle rain may spread over a wide area, but severe storms, which cause the chief problem, are almost always small in extent. There would be considerable benefit, therefore, in being able to switch the transmission path between earth stations some miles apart. Stations using two antennas that are several miles apart, linked to the same control center, are called *diversity* earth stations. They may become the normal form of earth station for systems using frequencies above 10 GHz. Possibly, in future networks, some transmission will not be "real time," as are telephone calls, but will include the transmission and *storage* of data, movies, facsimile mail, and so on. If this is the case, the duplexing over earth stations to avoid storms will not appear too wasteful, and the drop in capacity caused by a storm will merely delay the non-real-time transmission.

Clouds and fog are likely to cover a much wider area than storms. The attenuation they cause is not quite so severe, as can be seen from Fig. 15.13, which shows the attenuation caused by cloud cover from 2 to 5 km high [1]. Again, the higher the frequency, the higher the attenuation. The attenuation is proportional to the liquid water content of the clouds. A very dense cloud may have 2.5 grams of water per cubic meter. Isolated cases have been reported of clouds with as high as 4 grams per cubic

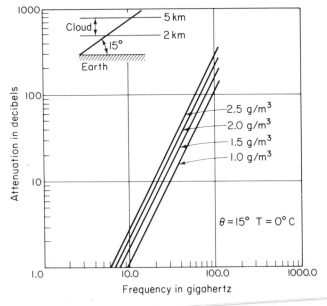

Figure 15.13.   The attenuation caused by clouds, in satellite transmission at different frequencies. These curves assume a uniform cloud cover extending from 2 to 5 kilometers in height. The angle of elevation is 15° and the temperature 0°C. The water content of the clouds are 1.0, 1.5, 2.0, and 2.5 grams/cubic meter. Only rarely do clouds have a water content higher than this. It will be seen that clouds affect millimeter wave transmission severely. *Reproduced from reference 1.*

meter, but this amount is rare. The angle of elevation of the transmission beams in the diagram is 15 degrees. Again large angles will be used for domestic transmission, and the attenuation will then be substantially less than that shown in Fig. 15.13. Ice clouds (surprisingly perhaps) give attenuations that are only a fraction of those of water clouds.

The air itself, and the water vapor in it, causes attenuation of millimeter waves, the absorption being higher at certain frequencies, as was shown in Fig. 13.14. If the frequencies of the absorption peaks are avoided, this attenuation is much less serious than that of rain and clouds.

Table 15.1 summarizes these three effects by showing percentile at-

Table 15.1 PERCENTILE ATTENUATIONS FOR RAIN, CLOUDS, AND AIR*

		Attenuation			
		Frequency (GHz)			
Percentile	Element	4	16	35	90
99.99	Rain (db)	0.5	26.4	105	320
(~ 1 hour	Clouds (2 g/m³, 0 °C) (db)	0.5	5.6	24	150
per year)	Oxygen and water vapor (db)	–	0.3	0.8	3.5
	Total (db)	1.0	32.3	129.8	573.5
99.9	Rain (db)	0.1	6.6	26.4	80
(~ 10 hours	Clouds (1.5 g/m³, 0 °C) (db)	0.3	4.0	17	120
per year)	Oxygen and water vapor (db)	–	0.3	0.8	3.5
	Total (db)	0.4	10.9	44.2	203.5
97	Rain (db)	–	0.4	1.5	4.7
(~ 12 days	Clouds (1.0 g/m³, 0 °C) (db)	0.2	2.7	12	82
per year)	Oxygen and water vapor (db)	–	0.3	0.8	3.5
	Total (db)	0.2	3.4	14.3	90.2

*Source: Adapted from reference 1.

tenuations for rain, clouds, and air, and the sum of these [1]. The top section of the table relates to the 99.99 percentile. In other words, the attenuation is worse than the figures shown for 0.01 % of the time— about one hour per year. The other two sections give the 99.9 percentile and the 97 percentile figures. The figures relate to a site with fairly heavy rainfall, 100 cm per year.

We see that except for 10 hours in the year, the attenuation of 16-GHz waves is not greater than 10.9 db and that of 35-GHz waves is not more than 44.2 db. The attenuation of 90-GHz waves is high, however. These frequencies have been selected because they do not coincide with atmospheric absorption peaks. The attenuation levels shown suggest that with large and powerful satellites, 16- and 35-GHz waves could be used successfully.

DIGITAL SATELLITES As will be discussed further in Chapter 17, a higher level of attenuation is possible with digital transmission than with analog transmission. The satellites discussed so far use analog amplifiers. The satellite of the future may receive an extremely high capacity bit stream, using millimeter waves, and then may regenerate the bits and retransmit them to earth. The bits stream will carry all types of signals, for example, in a *pulse code modulation* form.

The satellite, in this case, contains one or many solid state digital repeaters. Rather than amplifying a signal, which means amplifying the noise received also, the electronics will detect the presence or absence of bits, possibly using error-correcting codes, and will then regenerate a new, clean, bit stream. Large-scale integration circuitry, because of its extremely small size, low-power requirements, and high reliability, has an important role to play in digital satellite repeaters (see Chapter 16.)

OUTAGES Satellites are prone to a number of unique types of outages. First, there are eclipses. When the Earth's shadow passes across the satellite, its solar batteries stop operating. These eclipses last for about an hour on 43 consecutive nights in the Spring and Fall. The satellite can carry backup batteries to insure continuous operation if the need for continuous operation justifies carrying the additional weight.

Another form of outage occurs when the sun passes directly behind the satellite. The sun, being of such a high temperature, is an extremely powerful noise source and so blots out transmission from the satellite. This outage lasts about 10 minutes on five consecutive days twice a year.

A third form of outage is sometimes caused by airplanes flying through the satellite beam. And, lastly, if a satellite should fail, it may take some time to replace by a new launching. When satellites become a major part of the domestic network, there will no doubt be alternate satellites for use when eclipses and failures occur.

A LARGE COMMON CARRIER SYSTEM By using millimeter waves and pulse code modulation, a wide variety of spectacular satellite schemes becomes a possibility. Larger rockets than those launching today's satellites will probably be used during the 1970s, including, presumably, Saturn V class propulsion systems. Such rockets would permit the positioning of highly stabilized platforms weighing several tons.

For point-to-point traffic, a U.S.A. domestic satellite system, which might be regarded as an extension of the Bell System, could be established having extremely large capacity. Studies of such systems have been conducted at the Bell Laboratories [4] and elsewhere. One such study resulted in a system design and model using 50 ground stations and 50 satellites operating in two millimeter wavebands, 20 and 30 GHz, each band being 4 GHz wide. This would not interfere with existing and future satellites using different frequencies. It was stressed in the design that frequency space and orbit space are "precious and limited resources which must be conserved" [4]. The system has a total capacity as high as 100 million voice circuits or equivalent (e.g., 100 thousand television circuits). Such a capacity is far beyond today's needs, but future requirements will be high if the uses of telecommunications discussed here spread rapidly, especially Picturephone, large numbers of channels on cable TV, wall-screen TV, wideband data transmission, and video business conferences.

The Bell Laboratories study concludes that such a system would be "feasible within the next decade" (i.e., by 1978), although initially the full number of satellites would not be launched, and that "assuming a favorable outcome from the propagation measurements, particularly a demonstration that common-carrier standards for reliability can be achieved with diversity ground stations (switching between ground stations to avoid the effect of bad storms), plans for such a system could begin immediately" [4]. Needless to say, there would be a lot of arguing about allocation and system ownership before any such system was built.

The high system capacity is achieved (1): because of the very high bandwidth permitted by the use of millimeter waves and (2): because each satellite can be linked to each ground station and vice versa. If they are N channels, in the bandwidth used, G ground stations, and S satellites, the total system capacity is $N \times G \times S$ channels.

DIGITAL TRANSMISSION IS USED In the design in question, the satellites use eight two-way channels of 630 megabits per second each. Each satellite communicates with 50 ground stations simultaneously, with highly directive multibeam antennas. Each satellite therefore needs $8 \times 50 = 400$ repeaters for the 630 megabit stream. The satellite transmits with a power of 2 watts in the 20-GHz band and receives from the ground via the 30-GHz band. Each 630-megabit channel can carry 10,000 voice channels using pulse code modulation, so the total capacity of each satellite is 8 × 50 × 10,000 = 4 million voice circuits, or equivalent. Such a satellite would weigh about 5 tons, would use solid state integrated microwave circuits, and would have to be stabilized in orbit to within 0.01 degree. A rocket configuration based on the Titan 3-C could place this weight in synchronous orbit. *A configuration based on the Saturn V could place a much heavier satellite there.*

Smaller satellites could be used to serve smaller numbers of ground stations. Table 15.2 summarizes the parameters of typical satellites for repeating the 630 megabit per second stream.

Table 15.2 PARAMETERS OF PROJECTED DIGITAL SATELLITES*

| | Number of Ground Stations Served | | | |
	8	16	32	50
Weight of satellite (lb)	1640	3280	6560	11,500
dc power (kw)	1.15	2.30	4.60	7.20
Number of repeaters (number of 630 mb/s channels)	64	128	256	400
Total number of bits through the satellite (gigabits)	40	80	160	252
Number of one-way equivalent voice circuits through the satellite (thousands)	640	1280	2560	4000
Number of one-way equivalent TV channels through the satellite	640	1280	2560	4000

*Source: Adapted from reference 4.

The positioning of the ground stations will be important if satellites are used to carry the bulk of common-carrier long-haul traffic. The system must be designed so that the load will be reasonably well balanced between the ground stations. Many of the stations will be close to major metropolitan areas. Switching must be designed so that faulty circuits can be bypassed, as with today's terrestrial network. With a suitable switching matrix, a substantial number of failed repeaters in the satellites can be tolerated without inacceptable system degradation. The switching will make the system adaptable to changing traffic patterns and will permit ground stations affected by severe storms to be bypassed. Because of the transmission delay, the switching will be organized so that not more than one satellite link is switched into any circuit.

The Bell study assumes that the land link to the earth stations will be over the helical waveguide pipes discussed in the previous chapter. Eight pipes may be used, each carrying 60 two-way bit streams of 630 megabits per second. This amount gives a total of 4 million voice circuits, or equivalent, to each ground station.

INTERNATIONAL
SATELLITE
ALLOCATION

If satellites are not permitted to approach closer than 1 degree—as is likely to be the case with transmission between 10 and 40 GHz—then 95 is about the maximum number of "stationary" satellites using the same frequency that could serve the continental United States, as was shown, approximately to scale, in Fig. 15.11.

The worldwide allocation of such satellites may well become a severe problem. The number that could be launched today seems very large, but the telecommunicating world of the future will probably find uses for this capacity. The United States presumably cannot be allowed to use all, or even the majority, of the 95 satellites. But how shall this valuable future resource be divided among nations? Will the division be proportional to population—one man, one share of a satellite? In the latter case, the resource would be badly underutilized. Should the allocation be proportional to a country's contribution to Intelsat? This solution would also be difficult because Russia has its own version of Intelsat. China will presumably want communication satellites, too, and China is not accepted in the United Nations nor recognized by the United States. These are problems that the future must solve.

NONSTATIONARY
SATELLITES

The number of usable satellites could be increased by launching them into synchronous orbits that are not exactly over the Equator. When a satellite travels in a synchronous orbit inclined at a small angle to the Equator, it no longer appears stationary in the sky but appears to move in a figure 8, as shown in Fig. 15.14. Several satellites could follow each other around the figure 8. Using many such figure 8s, a large number of satellites could be launched, each permanently in view of an earth station and none of them approaching within 1 degree of another. A study at Bell Laboratories evaluated many such orbit schemes and proposed one for serving North America with a total of 477 satellites [5].

Such a scheme would probably still be controversial, internationally, because it increases the cost of the earth stations. Their antennas must now be able to track the satellites, and the minimum angle of elevation of the satellite decreases.

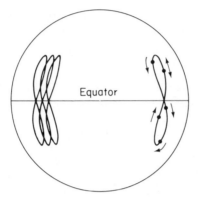

Figure 15.14. Satellites in a circular synchronous orbit inclined with respect to the equator appear to move in a figure 8. Packing schemes using multiple 8s would permit a much larger number of satellites than if they were all in an equatorial orbit.

NATIONAL SYSTEMS

There can be little doubt about the future value of satellite technology. Sooner or later satellites for domestic use within the United States will be as vital as the Bell System is today. Other nations and national groups appreciate this fact in varying degrees and are concerned as to how they can achieve their own national satellite systems. They have two choices; either they build their own launch capabilities or they use those of the United States or Russia.

Several nations now have the capability to place a small satellite in orbit, but it will be a long time before they are able to position a 5- or 10-ton satellite exactly, as the United States or Russia can. There is considerable concern about being dependent on these two countries for something as vital as the nation's communication facilities. A paper by a British Government official at the 1969 European Space Symposium [6] typically advocated the development by ELDO (the European Launcher Development Organization) of a suitable vehicle, Europa III. The paper stated that it was "not clear that a U.S. launching would be provided without prohibitive conditions." The schemes, however, that can be based on Europa III are pale by comparison with those discussed on the previous pages [6, 7, 8]. The United States may reap great commercial benefit from its large expenditure on space technology.

REFERENCES

1. "Future Communications Systems via Satellites Utilizing Low-Cost Earth Stations," prepared by the Electronic Industries Association for the President's Task Force on Communications Policy, Washington, D.C., 1969.
2. Eugene V. Rostow, "A Survey of Telecommunications Technology," The

President's Task Force on Communications Policy, Staff Paper No. 1, Washington, D.C., 1969.

3. "Communication Satellite Technology in the Early 1970s," Appendix C of Staff Paper No. 1 of the President's Task Force of Communications Policy, Washington, D.C., 1969.

4. Leroy C. Tillotson, "A Model of a Domestic Satellite Communication System," *Bell System Tech. J.*, December 1968.

5. Harrison E. Rowe and Arno A. Penzias, "Efficient Spacing of Synchronous Communication Satellites," *Bell System Tech. J.*, December 1968.

6. N. Simmonds, "Some Possible Concepts for a Second Phase of Communication Satellite Developments in Europe," European Space Symposium, London, 1969.

7. A. H. Schendal and J. Bouvet, "The Communications Satellite Project SYMPHONIE," European Space Symposium, London, 1969.

8. P. Blassel, "A European Telecommunications Satellite Project," European Space Symposium, London, 1969.

Valve panel at the first high-power radio station to use thermionic vacuum tubes for transmitting, 1923. Courtesy The Marconi Company Ltd., England.

16 CHANGING COMPUTER TECHNOLOGY

We have discussed the dramatic rise in transmission bandwidths and the wide-ranging capabilities of satellites. An equally important part of this story, some would say more important, is the changing nature of logic circuitry and computer storage.

Two decades ago logic circuitry implied the use of thermionic vacuum tubes. These tubes were large, unreliable, heat dissipating, and expensive. No replacement for the vacuum tube was in sight. In the late 1940s many informed and otherwise forward-looking scientists believed that the computer was a practical impossibility. Building a useful computer necessitated large numbers of logic circuits. Logic circuits needed thermionic vacuum tubes and vacuum tubes had a fairly high failure rate. A radio set with 5 vacuum tubes might fail only once per year, but a computer with 20,000 tubes would fail once an hour. Finding the failure would take more than an hour, and during that hour the machine would have to be kept running. Thus by the time one failure had been located, the chances are that another one would have occurred. The repair man would be chasing his own tail.

Now on my desk I have a circuit with over a thousand transistors, a thousand resistors, and many diodes. It is so small that it could almost fall through the hole in a punched card. It is also reliable, takes very little power, and can be mass-produced by a printing process that can produce it cheaply if made in large enough quantities. Such is the progress of two decades of electronics.

In 1948, before the transistor was invented, the possibility of miniature logic circuits, like those in today's missiles, would have been regarded as

the wildest science fiction. The next twenty years will bring equally startling changes, and we can be sure that some of the technology of 1990
would have been dismissed by today's engineers as "impossible." New
technology frequently comes from an unexpected quarter.

The significance of cheap, reliable, mass-produced logic circuitry is
twofold. First, there is data processing technology, extending as it will
into almost every walk of life. Second, in telecommunications, the transmission and switching of signals will be done in a digital rather than an
analog fashion. If inexpensive and reliable logic circuitry is available, the
telecommunications industry must rethink its use of multiplexers, concentrators, repeaters, switching offices, storage of signals, compandors,
and processing of signals to make them more compact. In the years ahead
we will see a merging of computer technology and telecommunications
technology.

MINIATURIZATION   The size of electronic circuits has been dropping by a factor of 10 approximately every
five years. Since 1950 it has dropped by more than 10,000 times. Figure
16.1 plots this progress. We are only beginning to understand the tech-

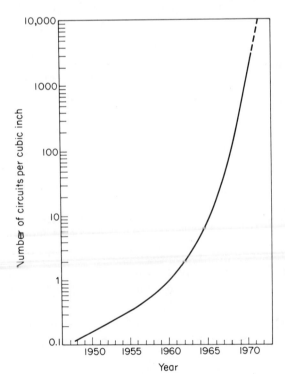

Figure 16.1. The increase in number of logic circuits that can be packed into a given volume. *Note that the vertical scale does not ascend in steps of equal magnitude. Each major division represents a tenfold increase.*

niques of large-scale integration; therefore it seems likely that the miniaturization process will continue at its exponential rate for at least another decade. The term *large-scale integration* (LSI) is ill defined but is generally reserved for circuits that carry out 100 or more individual functions (100 or more *gates*) and that are manufactured as a whole, having a density greater than about 50,000 components to the square inch. When the term *microelectronics* was first used, a chip of silicon one tenth of an inch square might have held about 10 transistors, together with many resistors, diodes, and capacitors. Now such chips can contain over 4000 transistors. Figure 16.3 shows photographs of such chips.

The transistors today are on a two-dimensional surface. One possible development during the next decade may be packing and connecting these surfaces together in a three-dimensional form. If the three-dimensional form could be accomplished with the density we have on a two-dimensional surface today, the density of components would be about a quarter of the density of nerve cells in the human brain. Eventually human-brain-packing density will probably be achieved, and the circuits will operate more than a million times faster than the brain on serial logic operations. A radical redesign of some basic computer concepts would be required, however, to permit entirely uniform arrays of such circuitry to be used.

Already the smallness of the circuits can permit previously unthinkable

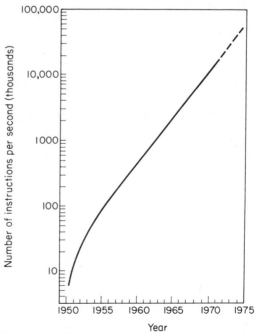

Figure 16.2. Increase in speeds of high-speed computers, in terms of memory cycles per second. *The verticle scale does not ascend in steps of equal magnitude. Each major division represents a tenfold increase.*

logic operations to be built into portable devices, such as pocket radio telephones or electronic devices in cars (see Chapter 20). The miniaturization and low-power requirements will be especially valuable in satellites.

INCREASE IN SPEED

Figure 16.2 shows the increase in speeds of high-speed computers. Their speed has been going up by a factor of 10 every seven years or so. There is a limit to the increase in speed of serial operations because electricity flows at a finite speed, approximately the velocity of light. In order to increase the circuit speed, the path length must become shorter. Hence speed is related to the degree of miniaturization. The exponential curve of Fig. 16.2 will probably continue for another decade due to large-scale integration. Ultimately the speed of electricity will impose a limit, but computers of the future may have a high degree of parallelism, which will again increase their overall speed. The Illiac IV computer project at the University of Illinois, among others, is exploring the prospects of highly parallel machines using LSI circuitry.

The very high bit rates of future telecommunications may also require the use of highly parallel LSI circuits in repeaters, multiplexers, concentrators, and time-division switching equipment. In the satellite that could handle 4 million voice channels (discussed in the previous chapter), many parallel repeaters would be used reconstructing digital bit streams, these bit streams being employed for all types of signals.

COST REDUCTION

The miniaturization and increase in speed are of great value, but perhaps more important is the reduction in cost that large-scale integration promises. Once a circuit mask is set up, the circuit can be mass-produced by a process somewhat similar to photographic printing, in vast quantities if so desired. The costs of this process depend on the number of steps involved and are largely unrelated to the complexity of the mask. Printing a photograph of a city is no more expensive than printing a photograph of a brick wall. For comparable production runs, it should cost little more to print 100,000 transistors on a chip the size of a postage stamp than to print 100.

On the other hand, the setup process is expensive. Preparing and debugging the mask are highly complicated. It is not economical to use LSI today unless several thousand of the one circuit are made, and the larger the production quantity, the cheaper the chips will be.

Where application can be found for them, some chips with a *very*

Figure 16.3   **LSI** (large-scale integration) chips **The** beginning of mass production in logic circuits.

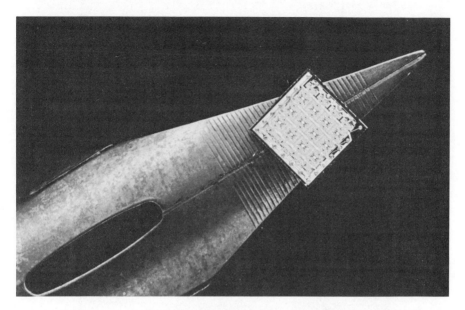

An IBM chip from the 360 model 195, balanced on a pen knife.

A ⅛-inch square chip from the Cogar corporation, with circuits containing 1500 transistors, shown on its mounting. Groups of these chips form the Cogar monolithic memory.

Figure 16.4. Because the design of complex LSI chips needs to be correct before they are manufactured (or tested), computer techniques are essential in their design. *Photograph courtesy of IBM.*

large number of components have been produced. The Bell System Picturephone camera, for example, has a silicon LSI target vidicon, something like the retina of the human eye. The target has 750,000 diodes within one-half inch square, each diode being 320 millionths of an inch in diameter.

Although chips have been made today containing several thousand circuits, these have not generally been used in logic units in computers. A chip with this number of circuits is difficult to design with assurance that it will contain no errors when manufactured and will not need to be modified after manufacture. The cost of changing and remaking the chip is extremely high. Furthermore, such a chip is exceedingly difficult to test —as is a complex program. It may have undetected bugs in it. Consequently, it is desirable to make chips with this number of circuits *uniform* in design—as is done with a memory array. In the near future 1000-circuit chips will be used in computers for memory but not for logic. Logic may be confined to chips of 100 circuits or less until means are devised for employing more uniform circuits for logic.

However, in general LSI use, the main problem now is how to take advantage of these circuits in such a way that a chip with many circuits can be mass-produced in the largest possible quantities. The result will probably be a radical redesign of the internal structure of digital computers. There may be a high degree of parallelism. Some startling schemes have been proposed for increasing the uniformity of the computer circuits. In one, for example, all memory "bits" have three possible states rather than the two states of today's memory; and this permits all manner of Boolean operations to be carried out in regularly structured memory rather than irregularly structured logic units. LSI is particularly advantageous in memory arrays, where large quantities of the same type of chip can be employed. It is also being used in machines that sell in large quantities, such as terminals.

Components used in telecommunications are likely to be needed in large quantities. There are 100 million telephone sets in the United States alone, and more than 10,000 public telephone exchanges. The use of LSI appears very attractive in pulse code modulation circuits, as well as in time-division multiplexing and time-division switching (discussed in the following three chapters).

The use of digital concentrators on local telephone loops could bring savings, particularly in locations where the subscribers are a long way from the central office.

Figure 16.5 shows the decrease in cost of digital circuitry, in terms of millions of computer instructions executed per dollar. This relates only to the cost of logic circuits, not to costs such as programming, input/output devices, or printers. In 1955 about 100,000 program instructions could be executed for one dollar. In 1960 the same dollar bought one million, and by 1970 it bought 100 million. In other words, it is going up by a factor of 10 every five years.

There are trade-offs in LSI circuitry between high speed, high packing density, and low cost. The fastest and most complex chips have a long manufacturing lead time, and the engineering changes so numerous on today's computers would be prohibitively expensive and slow. The designer, in other words, has to be correct the first time. Thus there is much to be said still for a small number of components per chips in the logic units. The engineering change problem may, again, give rise to radical structural changes in computer design.

The vertical scale in Fig. 16.5 and other figures in this chapter is a

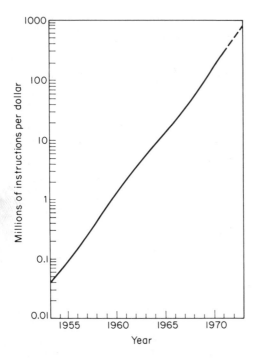

Figure 16.5 The decrease in cost of logic circuitry in terms of millions of instructions per dollar. *Note that the vertical scale does not go up in steps of equal magnitude. Each major division represents a tenfold increase,*

logarithmic scale, not a linear scale. In other words, a straight line represents a constantly increasing *rate* of growth, not a constant growth. This may be deceptive in that it might tend to give too low an idea of the growth. Suppose that we had drawn Fig. 16.3 with a linear scale instead, and suppose that we had drawn it so that the entire scale prior to 1955 had occupied *one inch* of the page. Then in order to show the improvement between 1955 and 1970, we would have needed a page 250 ft in height! My publisher would not permit this, and so we have had to use the logarithmic scale.

The improvement in cost in Fig. 16.5 is a truly outstanding one by normal accountancy standards, and lest the diagram should fail to illustrate this point, let us translate it into terms of personal finances. Suppose that you were able to invest money so that it had the same growth rate as in Fig. 16.5, and suppose also that one dollar had been invested for you at birth. By the age of 30 you would have been a millionaire (before capital gains tax.) If you had lived to be 70, then you would have overtaken the American Gross National Product!

RELIABILITY   The number of components in a computer has been increasing at a rapid rate. Unless the reliability of individual components increases at an equal rate, the reliability of the computer as a whole will drop. In fact, component reliability has been increasing by leaps and bounds, as is illustrated in Fig. 16.6. LSI circuits promise to be 100 times more reliable than the *integrated circuits* (IC) or *solid logic technology* (SLT) in general use today.

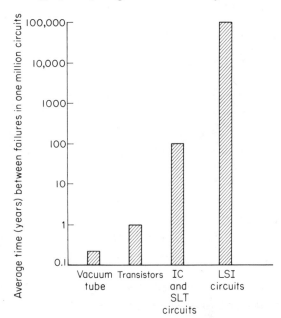

Figure 16.6. The increasing circuit reliability is essential as computers use increasing quantities of circuits. *Redrawn with permission from reference 1.*

In addition to the intrinsic reliability of this form of structure, advantage can be taken of the fact that extra components can be included on the chip at little extra cost, and self-correcting circuits can be printed. In memory arrays, for example, self-correcting codes can be used. If an error in a word or character is detected, the circuitry automatically corrects it. Instead of the parity checks most common today, self-correcting codes can also be used in shift registers, arithmetic units, and other components.

In some cases, the circuitry on the chip may be duplicated with automatic switchover on the chip in case of a component failing. Alternatively, two identical chips may be used in the design, thus taking advantage of the economics of quantity. This built-in reliability will be desirable in telecommunications equipment at unmanned locations.

In any installation it is desirable to minimize the maintenance calls, especially unscheduled ones. Self-correcting codes and automatic switchover features can help to do so. The device may register a fault that is occurring and being bypassed but still keep functioning correctly until its next scheduled maintenance.

BULK
STORAGE

LSI techniques will do for the computer industry what mass-production lines did for the automobile, only perhaps much more dramatically. LSI, however, is only one of the changes transforming computer technology. The capacity of bulk storage on computers is increasing at a faster rate than any of the curves in Figs. 16.1, 16.3, or 16.5.

Huge files of machine-processable information can be stored in two ways, "on-line" or "off-line." *On-line storage* means that the computer can read data without human intervention because the data are in the machine's own file units. The computer can read any piece of information at random, and it usually takes less than half a second to obtain it. *Off-line storage* refers to a medium such as magnetic tape to which the machine does not have direct access. Here the operator must take a tape from the shelves of a store room and load it; then the computer scans through the tape sequentially and thus relatively slowly.

Figure 16.7 refers to the amount of data the computer can have *on line*. It does not show the absolute maximum that one *could* have stored on line at any given time but rather a reasonable practical maximum that might have been found on an ambitious system at the time (not including military systems, CIA, NSA, etc.) Again the vertical scale is logarithmic.

As capacities go up, costs per character stored come down, as shown in Fig. 16.8. A *character* refers to one letter or digit or to a special symbol, such as a comma, dash, or question mark. This book, for example, contains about 200,000 characters (not counting the pictures). If it were on-line in a large file using the redundant coding that you are now reading, its proportion of the file cost would have been about $500 in 1955, $200 in 1960, about $2 in 1970, and perhaps only 2¢ in 1980. At the time of writing, systems exist in the laboratories which could store more than a million such books on line (more than 20 miles of library shelving). With efficient nonredundant coding, vastly more information could be stored.

The rate of cost reduction in Fig. 16.8 is greater than that in Fig. 16.5. Using the same illustration of investing a dollar at birth, with this growth rate you would overtake the Gross National Product by middle age, when you are still young enough to enjoy spending it!

The reader might again like to extend the lines of Figs. 16.7 and 16.8 upward for a decade and reflect on what could be achieved. It is staggeringly impressive (especially perhaps to a systems analyst who still thinks in terms of loading a file from punched cards). Is this exponential growth rate *likely* to continue? We cannot be sure. Certainly there must be an

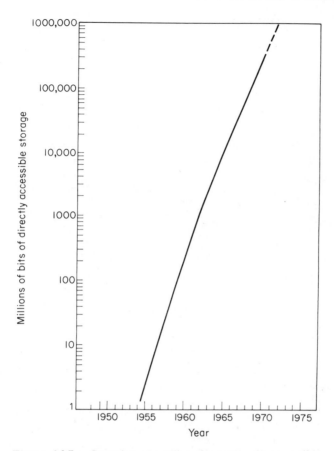

Figure 16.7. Capacity of on-line files, directly accessible by the computer. The curve relates to the maximum amount of storage likely to be found on large systems in each decade. *Note that the verticle scale does not ascend in steps of equal magnitude. Each major division represents a tenfold increase.*

upper limit somewhere. However, the developments now going on in the laboratories suggest that we are a long way from it yet. Before long we will be regarding today's billion character storage systems as abominably crude. And already the machines we are building are outstripping our ability to use them to good effect.

Terminals and telecommunications will be essential in order to load such files. Data *must* be captured at its source as in an airline reservation system or a terminal system for entering customer orders, production control data, and so on. During the 1970s vast data banks for a wide variety of applications will grow up—police information, legal documents, medi-

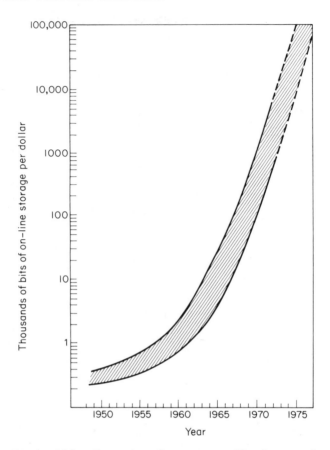

Figure 16.8. Cost of on-line storage. The increase in number of bits per dollar cost of on-line storage units. *Note that the verticle scale does not ascend in steps of equal magnitude. Each major division represents a tenfold increase.*

cal diagnostic data, market research data, medical case histories, and numerous others.

Because of LSI circuitry, mini-computers will be sold in great quantities. Much use of data transmission today is to obtain computing power on time-sharing computers. With mass-produced mini-computers, there will be less need to dial up computing power. Many people who need to do elaborate calculations will have a small machine in their office or at least in their building. Remote access to computers will often be to obtain *data* rather than to obtain computing power. The number of types of data, however, will be immense. Future uses of data banks far outstrip today's imagination, just as uses of computers entirely failed to be foreseen in the 1950s. With the cost of data storage dropping so rapidly, it will become uneconomic to store on paper in filing cabinets any records other than those of a very specialized or personal nature, such as hand-written letters, and even these will become machine-storable in image form. Via their terminal, the manager, the scientist, the doctor, the house-holder, the lawyer, or the schoolgirl will have enormous amounts of in-formation available to them if they know how to retrieve and use it.

Although the bit capacity of on-line storage will become prodigious, our usage of it and demand for it will probably grow equally fast. Many organizations have been surprised at the quantity of on-line storage their real-time schemes have consumed. In one system for keeping customer-order information on-line, the file requirements grew so large that it be-came impossible to have a two-dimensional machine floor, for the cable lengths available to the file units would not reach that far. Two floors had to be used with vertical cables! There are good reasons for storing some documents in image form rather than by coding the characters, but the storage required for this operation is high indeed. Image storage will nevertheless probably become a major part of data processing operations.

In fact, any signal that can be transmitted over telecommunication lines can be stored on files. We have already discussed the storage of voice on direct-access files. The storage of pictures and perhaps Picturephone images will follow. When cable TV or other terminal systems use still pictures, as was discussed in Chapter 5, these stills will be stored in some type of file in which they are randomly retrievable. Whether the storage will be photographic in nature and whether it will be digital or analog, future technology will decide. There will probably be extensive use of files that can be read by a computer but not updated by it. New photo-graphic processes and techniques using the laser and holography promise enormous capacity for such files. Perhaps a music library accessible by telecommunications will store its symphonies in holographic form. If advertising catalogs, available at terminals, have film sequences in them, the demand for on-line storage will be immense.

THE CHANGING COMPUTER WORLD

One of the exciting aspects of the computer world is that you can never be quite sure which directions the technology will take next. Com-puter experts have been taken by surprise several times by entirely new twists in machine design or application. Analog computer specialists were surprised by the overwhelming success of digital computers. Per-sons who saw the early digital computers as machines for scientific pur-poses were surprised when commercial applications swallowed most of the market. The coming of compilers and higher-level languages brought an unforeseen change. Magnetic tape in its day changed the realms of application possibility; then, to a world of batch processing, random access came as a new twist. At first disks were regarded by many people as a dubious gimmick, but they soon became essential to most systems and "data banks" became a fashionable term. In the 1960s data transmission became a major market, and, again, at first only the most imaginative computer users grasped its potentials. Real-time systems were another unexpected turn. Regarded initially as a highly specialized market for a small minority, they were soon to sweep the industry. Time sharing was considered a gimmick with intolerable overhead but boomed rapidly into a mass market.

It is almost certain that there will be more such "surprises." Technology in the laboratory today, dimly grasped or as yet unknown, will sweep into prominence. We may see a mass market for voice communication *to* as well as *from* computers. Associative storage may become a major factor. We will probably find much higher levels of automated programming. It may become common for programmers to sit at a screen unit and interactively enter their code, examining other persons' record layouts and being interrogated for their own documentation. Image processing may become a mass market. Terminals of the future may display pictures from computer-controlled microfilm storage, as well as digital output. A variety of techniques now classed as "artificial intelligence" may develop into maturity. We may find machines with a very high level of parallelism. We will almost certainly see new techniques for the control and organization of massive data banks. But probably the most important changes will be ones I have failed to foresee.

OPTICAL INFORMATION PROCESSING

Another twist that promises to become of great importance is *optical information processing*. This is as different from digital data processing as digital is to analog, and it can handle a whole new range of problems.

Optical processors are a class of machines that use lenses, mirrors, and masks to perform transformations on beams of light. The laser provides an intense source of coherent light for such machines. A wide variety of different mathematical transforms can be applied to the optical images. The entire image is manipulated at one instant and so makes possible a range of applications that would be difficult to handle by manipulating one or several serial bit streams, as in today's digital processing. The optical images can represent many different types of information. In an electronic signal, a *one-dimensional* quantity varies with time. Today's technology has many methods for manipulating this. An optical signal is one in which a *two-dimensional* quantity or image varies with time. Hence optical systems are capable of highly parallel processing. Some of the mathematical operations that optical systems can easily perform would be exceedingly complicated with digital processing.

Typical areas in which optical processing is likely to be useful are those involving pattern recognition, associative processing and areas in which a very large number of different data elements must be correlated. These will include print reading, fingerprint identification, the recognition of visual objects or faces, photographic interpretation, interpretation of geological or geophysical data, radar processing, image enhancement, secure communications, graphic information retrieval, voice recognition, and so on. In voice recognition, the one-dimensional time function— sound—would be converted into two-dimensional images for optical processing. A "voice-print" could thus become an image which needs to be identified like a fingerprint, and separate words can become separate images. Other time functions can similarly be processed as images.

A hologram contains information about two beams of light interfering coherently. If illuminated by a beam containing one set of information, the other set can be extracted. This process gives an extremely elaborate form of table look-up.

Much research is being done on optical information processing because of its military and intelligence applications. It is being used for processing aerial and satellite photographs. Changes in areas of interest such as in potential missile site can be highlighted, for example. Images can be enhanced in a variety of ways. It is used for processing radar data and for canceling the jamming of signals; and has many other potential uses.

Optical processing units may become an integral part of some data processing systems in the future. They may be used for telecommunication signal processing. In general, they open up an exciting variety of new uses for computing.

REFERENCES

1. Robert E. Markle, "Large-Scale Integration (LSI)—The New Technology and the Computer Community," AMA 16th Annual EDP Conference, 1970.

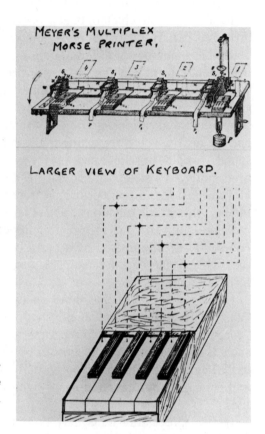

One of the earliest users of multiplexing—Meyer's Multiplex Morse Printer. Courtesy The Post Office, England.

17   PULSE CODE MODULATION

As mentioned previously, there are economic advantages to be gained from sending analog signals in a digital form on most telecommunication plant. The technological trends foreseeable are rapidly pushing the economics further in favor of digital transmission and time-division multiplexing. As these trends develop, the telephone networks will become, in some aspects, like a vast digit computer. This chapter discusses these trends in more detail.

THE BELL SYSTEM
T1 CARRIER

Chapter 6 explained the essentials of time-division multiplexing and pulse code modulation and illustrated them in Figs. 6.2, 6.3, and 6.4.

The most common use of pulse code modulation at the time of writing is in the Bell System T1 carrier. This highly successful system is leading to a hierarchy of carriers that will clearly dominate short-haul, and probably long-haul, transmission in the years ahead.

The Bell T1 PCM System multiplexes together 24 voice channels. Seven bits are used for coding each sample. The system is designed to transmit voice frequencies up to 4 kHz, and therefore 8000 samples per second are needed. 8000 frames per second travel down the line. Each frame, then, takes 125 microseconds. A frame is illustrated in Fig. 17.1. It contains eight bits for each channel. The eighth is used for supervisory reasons and signaling, for example, to establish a connection and to terminate a call. There are a total of 193 bits in each frame, and so the T1 line operates at $193 \times 8000 = 1{,}544{,}000$ bits per second.

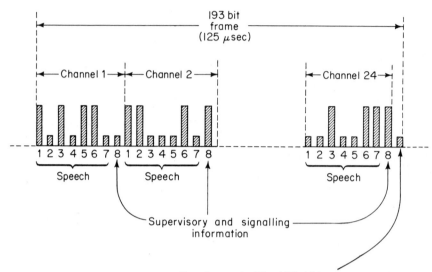

Framing code. The 193rd bits on successive frames follow a pattern which is checked to ensure synchronization has been maintained. If synchronization is lost this pattern is scanned for to re-establish it.

Figure 17.1. The bit pattern used to multiplex 24 voice channels on the Bell T1 system. *Redrawn from Davis, C. G., "An Experimental Pulse Code Modulation for Short-Haul Transmission." Bell System Tech. J. (January 1962).*

The last bit in the frame, the 193rd bit, is used for establishing and maintaining synchronization. The sequence of these 193 bits from separate frames is examined by the logic of the receiving terminal. If this sequence does not follow a given coded pattern, then the terminal detects that synchronization has been lost. If synchronization does slip, then the bits examined will in fact be bits from the channels—probably speech bits—and will not exhibit the required pattern. There is a chance that these bits will form a pattern similar to the pattern being sought. The synchronization pattern must therefore be chosen so that it is unlikely that it will occur by chance. If the 193rd bit was made to be always a 1 or always a 0, this *could* occur by chance in the voice signal. It was found that an alternating bit pattern, 0 1 0 1 0 1 . . . never occurs for long in any bit position. Such a pattern would imply a 4-kHz component in the signal, and the input filters used would not pass this. Therefore the 193rd bit transmitted is made alternately a 1 and a 0. The receiving terminal inspects it to ensure that this 1 0 1 0 1 0 . . . pattern is present. If it is not, then it examines the other bit positions that are 193 bits apart until a 1 0 1 0 1 0 . . . pattern is found. It then assumes that these are the framing pulses.

This ingenious scheme works very well with speech transmission. If synchronization is lost, the framing circuit takes 0.4 to 6 milliseconds to detect the fact. The time required to reframe will be about 50 milliseconds at worst if all the other 192 positions are examined; but normally the time will be much less, depending on how far out of synchronization it is. This is quite acceptable on a speech channel. It is more of a nuisance when data are sent over the channel and would necessitate the retransmission of blocks of data. Retransmission is required on most data transmission, however, as a means of correcting errors that are caused by noise on the line and detected with error-detecting codes.

COMPANDING The quantization of the signal needed for pulse code modulation is not linear as in Fig. 6.3. If the signal being transmitted were of low amplitude, then the procedure illustrated in the figure would not, of itself, be so satisfactory. The quantizing noise, still the same absolute magnitude, would now be larger relative to the signal magnitude. The quantizing error is a function of the interval between levels and not of the signal amplitude; thus the signal-to-quantizing-noise ratio is lower for smaller signals. For this reason a compandor is normally used to compress (and later expand) the very wide dynamic range of speech signal amplitudes to minimize this effect.

A *compandor* is a device that, in effect, compresses the higher-amplitude parts of a signal before modulation and expands them back to normal again after demodulation. Preferential treatment is therefore given to the weaker parts of a signal. The weaker signals traverse more quantum steps than they would do otherwise, and hence the quantizing error is less. This is done at the expense of the higher-amplitude parts of the signal, for the latter cover fewer quantum steps.

The process is illustrated in Fig. 17.2. The effect of companding to move the possible sampling levels closer together at the lower-amplitude signal values, is sketched on the right-hand side of the figure, which shows the quantizing of a weak signal and a strong signal. The right-hand side of the diagram is with companding, the left-hand side without. It will be seen that on the left-hand side the ratio of signal strength to quantizing error is poor for the weak signal. The ratio is better on the right-hand side. Furthermore, the strong signal is not impaired greatly by the use of

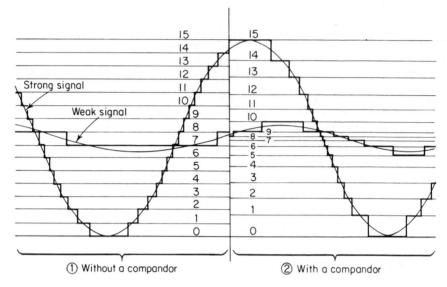

Figure 17.2. With a compandor the quantization of the weak signal gives more separate values, and therefore a better ratio of signal to quantizing noise.

the compandor. In practice, the PAM pulses are companded and one equipment serves all the channels, which are being multiplexed together.

A more modern scheme uses a coder with built-in nonlinear quantization.

REGENERATIVE REPEATERS
The main reason why high bit rates can be achieved on wire-pair circuits using pulse code modulation is that repeaters are placed at frequent intervals to reconstruct the signal.

In most PCM systems working today the repeaters are placed at intervals of about a mile. The Bell T1 System, operational since 1962, uses repeaters at intervals of 6000 ft, which is the spacing of loading coils employed when the wires were used for analog transmission; the repeaters replace the loading coils. These repeaters reconstruct 1,544,000 pulses per second. A higher bandwidth coaxial link was operated experimentally by the Bell Laboratories in 1965;[2,3] it transmits 244 million pulses per second with repeaters every mile. A ten-mile link had an error rate below 10^{-10}, and this rate will be good enough for a 4000-mile link coast to coast in the United States.

A regenerative repeater has to perform three functions, sometimes referred to as the 3 "R"s: reshaping, retiming, and regeneration. When a pulse arrives at the repeater, it is attenuated and distorted. It must first pass through a preamplifier and equalizer to reshape it for the detection process. A timing recovery circuit provides a signal to sample the pulse at the optimum point and decide whether it is a 1 or 0 bit. The timing circuit controls the regeneration of the outgoing pulse and ensures that it is sent at the correct time and is of the correct width.

HIERARCHY OF DIGITAL CHANNELS   As pulse code modulation systems increasingly replace frequency-division multiplexing, a hierarchy of digital facilities will be built. AT & T has plans for a nationwide network of PCM channels using wire-pair, coaxial cable, and microwave transmission. On this network, voice, television, facsimile, Picturephone, and computer data will all travel intermixed.

The T1 system will be part of the network. The other facilities planned are shown in Fig. 17.3. The T1 system carries a "channel bank" of 24 voice channels with time-division multiplexing. It transmits 1.544 mil-

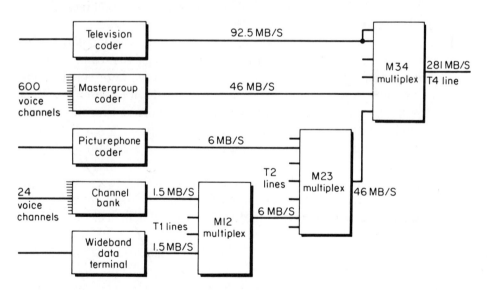

In the proposed digital hierarchy, signals will be multiplexed in several stages. For example, four T1 signals will be combined into a 6 megabit T2 signal. Seven T2 signals then will be multiplexed, forming a 46 megabit pulse stream. This intermediate rate matches that of a coded mastergroup. Six 46 megabit signals will be multiplexed, forming a 281 megabit T4 signal. Other combinations are possible. For example, two television and two mastergroup signals form a T4 signal.

Figure 17.3. Redrawn and reproduced by permission of the editor, Bell Laboratories Record.

lion bits per second, and a wideband data terminal can be connected to it. The system, now coming into widespread use, must be a component of any larger time-division multiplexing hierarchy.

The next step is the T2 line, which carries a pulse stream of 6.3 million bits per second. An M12 multiplex unit combines the bits from four T1 lines to travel over one T2 line. The T2 line is of the capacity required for one Picturephone signal. The Picturephone image when digitized needs about 6 million bits per second. If AT & T meets its predicted sales of Picturephone, there must be a major network of such lines and switching centers by 1980. Many of these facilities will be in operation in the early 1970s.

Seven T2 lines, carrying voice data, Picturephone, or any other signals, can feed into an M23 multiplex unit. Their signals are interleaved to form a signal at 46 million bits per second. It is planned that today's master group can be digitized as a unit to form a signal of 46 million bits per second. Color television needs twice this bit rate and could be carried on two such pulse streams.

At the next higher level, the M34 multiplex unit combines six 46-megabit signals to form a signal at 281 million bits per second. This signal would travel over a T4 link. At present, there is no transmission facility planned between the T2 and T4 to carry the 46-megabit stream.

The T4 or T2 pulse stream will travel over microwave or coaxial cable links similar to those in use today but redesigned with digital repeaters. Such a means of transmission seems well suited to the new media—helical waveguides and laser beams; and AT & T has proposed a satellite system that employs digital transmission and is interconnectable with the preceding network.

One great advantage of such a system is the ability to carry all types of signals together without the problem of having them interfere with each other. The terminals and multiplexing equipment will probably be built at lower cost than today's equipment via the large-scale integration technology. A characteristic of LSI circuitry is that it becomes low in cost if a sufficiently large quantity can be made to offset the initial design and production of the masks. This factor is certainly likely to be the case with these types of links. Before digital systems are introduced, they must show improvements in cost and performance over existing technology. The Bell T1 system has shown such improvement. It seems probable that the T2 and T4 systems may do so also, and it is widely expected that such systems will eventually predominate. The prospect is an exciting one for the computer world.

SYNCHRONIZATION   Synchronization is clearly important with digital transmission. It is essential for the receiving machine to know which bit is which. This is not too much of a problem on point-to-point lines. The Bell T1 carrier solved it by adding one extra bit per frame, thereby obtaining an 8000-bit-per-second signal that carries a distinctive pattern. If synchronization slips, then this pattern is searched for and synchronization can usually be restored in a few milliseconds.

This is fine so long as the channel bank remains intact. If, however, the bank were split up into its constituent channels and these channels were transmitted separately by pulse code modulation, then it would be advantageous to have a synchronization bit sequence *for each channel* rather than for the group of channels. The result would be one bit per character rather than one bit per frame, as in Fig. 17.1. One bit per character is already used for network control signaling, and the result is 8000 bits per second in this case. This rate seems far too much for any purposes that can be foreseen at the moment. Network signaling consists mainly of sending routing addresses (the number you dial) and disconnect signals (when you replace your receiver). Therefore it has been suggested that this bit position should be shared between the network control signaling function and the synchronization function. As a result, it would take twice as long to reachieve synchronization than if the bit per character were used solely for signaling. Still for an individual channel by itself, the time does not seem excessive.

It seems likely that the above approach may be recommended as a world standard by CCITT, the Geneva-based organization controlled by the United Nations, which has done so much to help standardize world telephony and telegraphy so far.

Synchronization becomes a much more difficult problem when a large switched network is considered. Signals are transmitted long distances over different and variable media, and their time scale must inevitably differ slightly even if the transmitting locations are synchronized. This problem must be solved before a nationwide PCM network can be established; consequently, various solutions are being worked out. These solutions fall into two categories. The first is a *fully synchronous* approach in which an attempt is made to synchronize the clocks of the different switching offices and to compensate for any drift in synchronization of the information transmitted. The clocks of the different offices in the network must all operate at the same speed. The second is a quasi-synchronous approach in which close but not perfect clock correlation is accepted. Because the clocks differ slightly in speed, certain minor mutilations of the information carried will occur. These may be overcome by adding dummy characters to the transmission and confining the mutilation as far as possible to the dummies. This technique is referred to as *pulse stuffing.*

Mutilations of the information occurring during resynchronization can do negligible harm to voice transmission. It does not matter if an occasional millisecond of information is lost now and then; with data, however, it can be more harmful. When mutilation of the intelligence occurs, retransmission of a block of data must follow. This must not occur too frequently. The synchronization needs of a network that will be used to transmit data are tighter than those for a network that only transmits speech. One would hope to achieve clock correlation of one part in 10^7. This figure, with a sampling rate of 8000 samples per second, means possible mutilation of a sample once every 1250 seconds, which is probably acceptable.

It is vital that in the formative stages of these networks the future needs of data processing be fully considered.

When transmitting analog signals such as speech or television, there are methods of digitizing the signal which can be more efficient than encoding the value of each sample as does PCM. It is possible to encode the change in voltage at each sample time rather than the absolute value of the voltage.

One such technique is referred to as *delta modulation,* and it is illustrated in Fig. 17.4. Here only one of two conditions is transmitted at each sampling time: *plus* or *minus.* The sample voltage is compared, each time, with a voltage obtained by integrating the previous samples. If it has gone up, a bit representing a *plus* condition is transmitted.

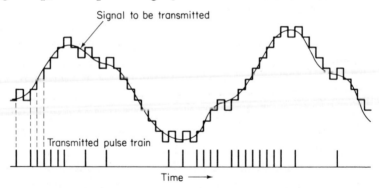

Figure 17.4. Delta modulation.

The number of pulses needed for this form of encoding depends on the rate of change of the signal amplitude. If the peak amplitudes are of low frequency and if the high-frequency components are of low amplitude, fewer bits will be needed for the encoding than if all frequency components are of the same amplitude. This is the case with speech. Figure 17.5 shows a typical speech spectrum. It is also the case with television and Picturephone signals. In a 1-MHz Picturephone signal, most of the energy is concentrated below 50 kHz.

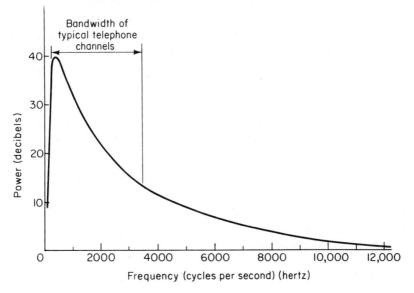

Figure 17.5. The spectrum of human voice. The high amplitudes are at low frequencies. This makes it particularly suitable for transmission by *delta modulation.*

Overloading with this type of modulation will come not from too great a signal amplitude, but from too great a rate of change. The encoding on the right-hand side of Fig. 17.4 is barely keeping up with the signal change.

It is inefficient to use only one bit per sample, and another variation on this scheme uses three or four bits per sample which permits eight and sixteen gradations of signal change to be recorded each sample time, and then uses fewer samples. The levels, as with the right-hand side of Fig. 17.2, are not linearly spaced.

It is possible that for transmitting speech or television some such scheme will prove to give better results than straight PCM.

There are several other more elaborate coding schemes in operation that can achieve excellent quality transmission using a lower bit-rate.

PICTUREPHONE ENCODING   A 1-MHz Picturephone signal is carried by a bit stream of 6.3 million bits per second. The sampling rate must be 2 million samples per second (twice the maximum frequency, as discussed in Chapter 6). If standard PCM encoding is used, this gives 3 bits per sample. Hence eight discrete amplitude levels can be reconstructed. This scheme results in a grainy picture. Eight levels are not enough to avoid grain. Color television is transmitted using 10 bits per sample.

To overcome this problem differential *encoding* is used, as is illustrated in Fig. 17.6. A feedback loop is employed and a signal which is delayed by a small amount is subtracted from the incoming signal. This produces a difference signal, as seen at the bottom of Fig. 17.6. When the object in the camera view is still, the incoming signal is not changing, and the difference signal is zero. For slight movement in front of the camera, the difference signal is small. For rapid movement, the difference signal is large. Most of the time the Picturephone image changes slowly, and so the eight sampling levels are bunched together around zero amplitude, as shown. Fewer levels are used for larger difference signals.

When the subject is moving rapidly, there is still some graininess with this scheme. However, the overall subjective effect is much better than with straight PCM encoding.

IN-PLANT SYSTEMS   The ability of pulse code modulation to give a high bit rate on a wire pair is also used in in-plant systems (systems using wires within a factory or office building that are private and not part of any common facility). Again the key to such systems is the placing of digital repeaters at sufficiently frequent intervals along the wire pair.

An example of this situation, which seems a pointer to future trends, is found in the IBM 2790 data collection system, illustrated in Fig. 17.7. Here a 22-gage twisted wire pair is used with digital repeaters every 1000 feet. Again the repeaters reconstruct the bits, which they pass on, and the effect of noise is not cumulative. A transmission speed of 500,000 bits per second is used and the wire pair can transmit this in an electrically noisy factory environment with a very low error rate. The wire pairs are arranged in closed loops from a transmission control unit that can be connected to a computer directly or remotely as in Fig. 17.7.

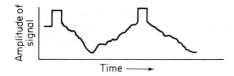

The original analog signal

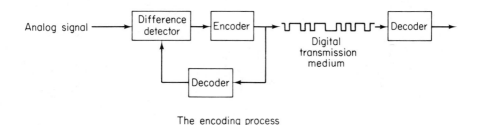

The encoding process

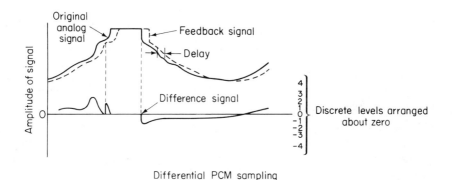

Differential PCM sampling

Figure 17.6. The differential encoding process in Picture-phone transmission. *Redrawn with permission from "Transmission Across Town or Across the Country" by D. W. Nast and I. Weller, Bell Laboratories Record, May/June 1969.*

One of the most interesting factors in the IBM 2790 System is that the bit stream on the wire loop could carry so much more data than it actually does. Without doing anything particularly ingenious, it could be made to carry 18 times as much information. At present, the 500,000 bits per second carry a maximum of 14,815 bits of information. The primary concern of the design team was to produce low-cost terminals so that an organization could afford to have numbers of them on the shop floor of

USE OF A HIGH-SPEED PULSE-CARRYING LOOP:

A simple pair of wires can be made to carry a very high bit rate if repeaters are used at sufficiently frequent intervals to regenerate the bits. This is the principle of the Bell system T 1 carrier, using pulse code modulation of 1,500,000 bits per second. It is also used on some in-plant systems to give a high bit rate on loops of wire around a factory or laboratory. The IBM 2790 system uses a 500,000 bit per second pulse stream on loops of No. 20 AWG twisted wire pair with repeaters every 1,000 feet:

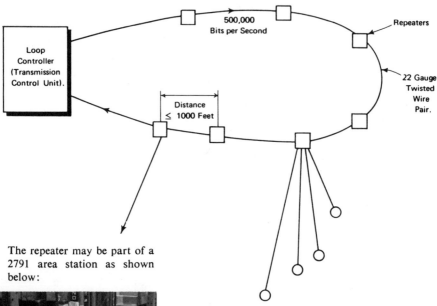

The repeater may be part of a 2791 area station as shown below:

Various devices may be attached to the area station on other wire pairs, for example, this 2796 data entry unit:

Figure 17.7

its factory. The new PCM technology provided more bits than necessary, and one of the slogans of the design team was "bits to burn."

If the computer were able to handle the work load, 800 terminals could be attached to today's IBM 2790 wire loop. However, with tighter organization of the data on the loop, at least 10,000 could be attached.

It is interesting to do calculations about the attachment of more complex terminals to such a loop. Consider visual display terminals of the type used in applications like airline reservations. Most applications use alphanumeric messages in their man-machine conversation. There are usually pauses between some of the messages while the operator thinks, talks to a client, or carries out some other function. Not all the terminals on a typical system are in use at the same time. Averaging over all the terminals, it is rare to find a mean data rate of greater than 100,000 bits per hour per terminal. This is substantially higher than the peak data rate on a typical airline system, for example. At this rate, a single wire pair could carry the traffic from more than 10,000 terminals—using synchronous transmission—without undue queuing problems.

PCM loops, perhaps using a facility like the T1 carrier, may well prove to be the best way to hook a large number of terminals to a computer.

**APPLICATION
OF PULSE CODE
MODULATION**
Today the main advantage of PCM to the common carriers is in short-haul voice transmission. Use of the T1 carrier for this purpose is becoming widespread in the United States. The reason is that there is a rapid rise in traffic over short distances, and present transmission media are fully utilized. By replacing the loading coils with digital repeaters on wire-pair lines, these lines can be made to carry far more traffic.

Over very short distances, multiplexing is not worthwhile. The cost of the multiplexing equipment offsets the increased utilization of the lines. In some cities, however, PCM will prove economical over distances of only a few miles. The existing cables and duct routes are often fully utilized, and the provision of additional ones requires digging up congested streets at great cost.

The conservation of bandwidth is more important over very long distances. As we have seen, fewer voice channels can be sent over a given bandwidth with time-division than with frequency-division multiplexing. The cost of the equipment for frequency-division multiplexing is much higher than with time-division multiplexing, but for long-distance transmission, there are few such equipments per mile of channel. In general, today both the quality and economics of long-haul transmission are more satisfactory than those of the tributary networks, and therefore there is less incentive to change them to PCM.

However, long-distance bandwidth is dropping in cost. The introduction of CLOAX cable, the waveguide, the domestic satellites, and eventually the laser will cause it to drop farther. At the same time, PCM costs are dropping. The synchronization problems of large networks are being solved; the T5 and T6 carriers are being designed for mass production. Furthermore, if the short-haul lines are PCM, it becomes economically advantageous to interconnect them with long-haul PCM lines. Before many years, new plant installed for both short-haul and long-haul telecommunication links is likely to be digital.

Figure 17.8 shows how PCM lines might be used in today's typical urban environment. It presents a highly simplified picture of part of the telephone network in London. The top half of the picture shows interconnections between local, tandem, and subtandem exchanges. The bottom half shows how the network could be simplified with the use of PCM links. The simplification shown by the illustration would have appeared far more striking if it had been drawn with the 200 or so local exchanges that exist rather than with the small number in the diagram.

The saving would be much greater, again, if PCM concentrators were used also, as will be described in the next chapter.

ADVANTAGES OF PCM

As noted earlier, PCM can give lower costs per telephone channel on short-haul lines and can greatly multiply the utilization of lines within city areas. Two trends will widen the range of economic application: the decrease in bandwidth cost due to the introduction of higher-capacity channels and the decrease in cost of logic circuitry, which is likely to be great when large-scale integration is fully developed.

PCM has several other advantages in addition to this overriding economic argument:

1. Because the repeaters *regenerate* the bit stream, PCM can accept high levels of line noise, cross talk, and distortion. A substantially worse noise-to-signal ratio can be accepted than with frequency-division multiplexing.

2. The transmission is largely unaffected by fluctuations in the medium, provided that they do not exceed certain limits. This is termed *ruggedness* in a transmission system.

3. All types of different signals, such as speech, music, television,

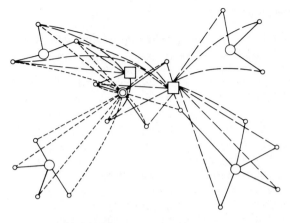

○	Local exchange
◯	Subtandem exchange
◎	Tandem exchange
▢	Toll/trunk outlet
——	Junction to subtandem
-----	Junction to tandem
—·—	Junction to toll/trunk outlet

A highly simplified diagram of the present London step–by–step tandem network catering for subscriber trunk dialing (direct distance dialing in American parlance). Only a limited number of exchanges and routings are shown, in order to lessen the complexity of the figure. All direct junctions between local exchanges have been omitted.

With integrated PCM between exchanges, the routing might be simplified as follows:

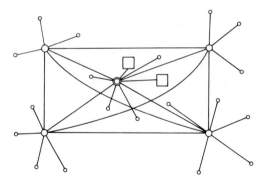

If the drawing had not been so highly simplified, the saving would have appeared much greater.

Figure 17.8. Redrawn with permission from Techniques of Pulse Code Modulation in Communication Networks, by C. C. Hartley, P. Morret, F. Ralph, and D. J. Tarran. Cambridge University Press, Cambridge, 1967.

Picturephone, facsimile, telegraphy and computer data, will be multiplexed together in a digital form. These signals can all travel together over the same facilities without interfering with one another. In analog channels the system capacity is often limited by mutual interference between different types of signals.

4. Much higher data rates can be achieved than with analog transmission. This factor will become increasingly important economically as the use of computers and terminals increases. Data transmission is increasing much faster than other forms of transmission. Unless held back by unsuitable or expensive transmission facilities, it will continue to rise probably at an increasing rate. Most common-carrier predictions of this market seem gross underestimates. There will be little relationship between present-day telegraphy and the future uses of data transmission.

5. For some future transmission media, such as laser pipes, digital transmission may prove to be the only practicable method.

6. Many bits are available for network control signals. The possibilities for signaling and remote control of the network are therefore much greater than with today's analog plant.

7. Encryption of signals is likely to become an important subject with the increasing concern about privacy and increasing need for security in data processing systems. Digital transmission makes effective encryption easy to achieve. Encrypting devices may perhaps be used in the private branch exchange of the future, under computer control. The analog scramblers, familiar to viewers of World War II movies, are of little use today because computer methods make decryption of the scrambled signals easy.

8. Time-division multiplexing provides advantageous switching methods as well as transmission multiplexing. The networks of the future will integrate switching and transmission technology, both, in part, using digital techniques. Time-division switching costs will be lower if the transmission also uses time-division multiplexing.

9. Concentrators can substantially lower the cost of the local distribution network. As will be discussed, digital techniques provide a way to build inexpensive concentrators. These devices could be used in large numbers in future telecommunication networks.

REFERENCES

1. J. S. Mayo, "A Bipolar Repeater for Pulse Code Signals" *Bell System Tech. J.*, January 1962.

2. J. S. Mayo, "Experimental 224 MG/S PCM Terminals" *Bell System Tech. J.*, November 1965.

3. I. Dorros, J. M. Sipress, and F. D. Waldhauer, "An Experimental 224 MG/S Digital Repeatered Line" *Bell System Tech. J.*, September 1966.

4. C. G. Davies, "An Experimental Pulse Code Modulation System for Short-Haul Trunks" *Bell System Tech. J.*, January 1962.

5. M. R. Aaron, "PCM Transmission in the Exchange Plant" *Bell System Tech. J.*, January 1962.

6. R. H. Shennum and J. R. Gray, "Performance Limitation of a Practical PCM Terminal" *Bell System Tech. J.*, January 1962.

7. H. Mann, H. M. Straube, and C. P. Villars, "A Companded Coder for an Experimental PCM Terminal" *Bell System Tech. J.*, January 1962.

8. K. E. Fultz and D. B. Penick, "The Tl Carrier System" *Bell System Tech. J.*, September 1965.

9. J. F. Travis and R. E. Yeager, "Wideband Data on T1 Carrier" *Bell System Tech. J.*, October 1965.

10. J. O. Edson and H. H. Henning, "Broadband Codes for an Experimental 244 MG/S PCM Terminal" *Bell System Tech. J.*, November 1965.

11. F. J. Witt, "An Experimental 224 MG/S Digital Multiplexer-Using Pulse Code Stuffing Synchronization" *Bell System Tech. J.*, November 1965.

A typical rural telephone exchange in the Netherlands (about 1930). As new telecommunication facilities come into operation, they have to coexist with the old. Courtesy Dutch Posts, Telegraph and Telephone Administration, The Hague.

18 TIME-DIVISION SWITCHING

When analog signals are decomposed into a set of samples, as shown in Fig. 6.2, it is possible to transmit many different signals together in a high-speed pulse stream by interleaving the samples. It also makes possible *time-division switching*.

In time-division switching, the different samples are sent down different paths, depending on their desired destination. Figure 18.1 illustrates the principle. Here tomatoes are shown rolling down a chute and being sorted by two gates that open and close at exactly the right times. There are three types of tomatoes: A, B, C, A, B, C, A, B, C, . . . They represent samples from three signals traveling together in a time-division-multiplexed fashion.

It is desirable to switch signal A on to path 2. The gate connecting path 1 with path 2 opens at exactly the right times to make the type A tomatoes roll down path 2. Similarly, signal C must go down path 3. The gate to path 3 opens at the right times to achieve this step. At other times it might be desirable to switch the A signal down path 3. This is accomplished by changing the timings with which the gates open.

In a time-division switch, the samples might be arriving at a rate of one million per second. There might be hundreds or thousands of paths, and the gates open and close at electronic speed like the logic circuits in a computer.

Path 1 in Fig. 18.1 may be thought of as being equivalent to an electrical *bus* that carries pulse amplitude modulated (PAM) samples (Fig. 6.2) from a group of lines. Outgoing lines must be electronically con-

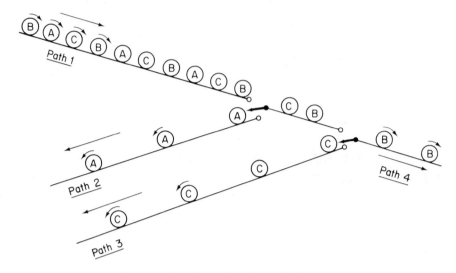

Figure 18.1.   Time-division switching.

nected to this bus at instants such that the requisite samples flow down
them.

CONCENTRATORS   Figure 18.2 illustrates the principle of a concentrator that uses time-division switching. In
this simplified picture, five telephones are connected to two outgoing
lines. All seven lines entering the switch carry analog voice signals. Each
goes through a low-pass filter that removes any frequencies higher than
4000 Hz; the lines are then connected to a bus via a sampling gate. Each
sampling gate operates 8000 times per second and thus produces 8000
PAM pulses per second (8000 is the number of samples needed to reconstruct completely a 4000-Hz signal).

Since there are seven lines, the bus must carry seven sets of samples—
that is, $7 \times 8000 = 56,000$ PAM pulses per second. The gate on any
line can be controlled so that it will open at times equivalent to any of
these sets of samples. If the telephone 1 is to be connected to outgoing
line number 2, then the switch on the line to telephone 1 and the switch
on outgoing line number 2 must be opened at the same instants.

Because the bus carries seven sets of pulses, this step could take place
at the same time as telephone 4 is connected to telephone 5, say. Computerlike logic opens and closes the gates at the correct instants. The
entire device would be built in solid-state circuitry.

The number of lines that could be connected to such a device depends on the number of PAM pulses the bus can carry. If the bus can carry a million PAM pulses per second, then $1,000,000/8 = 125,000$ lines can be connected to it.

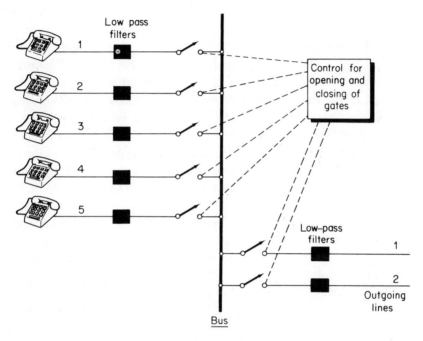

Figure 18.2

It will be seen that time-division switching is quite different from conventional switching, in which a physical path is permanently connected (called *space-division switching*). The number of switch points is much lower. If there are N lines, then N switch points are needed. If, on the other hand, N lines are to be interconnected physically, then N^2 switch points are needed. The latter figure can be reduced by multistage switching with a limited number of simultaneous interconnections; if no more than one tenth of the lines are permitted to be interconnected at one time, $0.21 \, N^2$ switch points could be used. Thus for a switch interconnecting 100 lines, time-division switching needs 100 switch points, whereas open space-division switching is likely to need at least 2100. Time-division switching is lower in cost today than space division for many applications. As the cost of fast logic circuitry drops lower, time-division switching will become increasingly economical.

ESS No. 101
SWITCH UNITS

An interesting example of time-division switching is found in the way subscribers are connected to the Bell No. 101 Electronic Switching System which will be discussed in the next chapter. A subscriber location might have, say, 200 telephone extensions. Any extension must be able to dial any other and also be able to dial numbers on the public network. This process is conventionally achieved with an electromechanical private branch exchange. In the No. 101 Electronic Switching System, the dialed digits are interpreted in a computer at the local central office, which then controls the switching. All that is required on the subscriber's premises is a small cabinet of electronics, which samples the signals on the voice lines and switches them in a time-division fashion.

The organization of the equipment in this cabinet is shown in Fig. 18.3. There are two buses onto which the samples of signals on the voice lines are switched. The speech is sampled 12,500 times per second. Thus there are 80 microseconds between sampling times. The duration of each sampling is approximately 2 microseconds, with a guard interval of 1.2 microseconds between samples. Therefore there can be 25 independent sets of samples or 50 for both buses, giving a total of 50 time slots to be divided among 200 speech lines. No more than a quarter of the extensions can be in operation at one time—a higher traffic-handling capacity than normally encountered on private branch exchanges of this size.

The buses thus carry $12,500 \times 25 = 312,500$ PAM samples per second each. Samples are gated to the appropriate speech lines under the control of electronic circuitry, which is directed by the computer at the local central office. For this purpose, two data lines go from the time-division switch unit to the central office computer. One carries the dialed or Touchtone digits; the other carries the switching instructions to the switch unit.

The use of two buses doubles the number of time slots available to the customers, but perhaps more important, it greatly enhances reliability. If one bus or its controls fails, the other can handle any of the extensions. In this way, 25 time slots instead of 50 are then available. The gate shown between the two buses also gives a convenient means for establishing conference calls. Figure 18.4 shows the bus arrangement of a larger switch unit. Here there are up to four buses. Each carries 60 independent time slots, giving a total of 240, which means 3 million samples per second. Subgroups of 32 speech lines are connected through the time-division switches to subgroup buses. The subgroup buses in turn are connected through more time-division switches to the 60-time-slot buses.

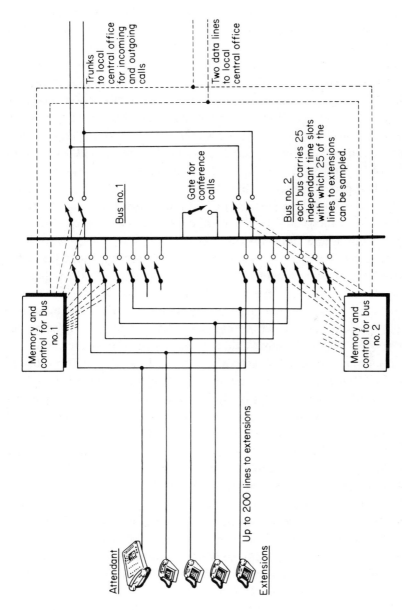

Trunks to local central office for incoming and outgoing calls

Two data lines to local central office

Bus no.1

Gate for conference calls

Bus no. 2
each bus carries 25 independent time slots with which 25 of the lines to extensions can be sampled.

Memory and control for bus no.1

Memory and control for bus no. 2

Attendant

Up to 200 lines to extensions

Extensions

Figure 18.3. The organization of the time-division switching units that are connected to the Bell System No. 101 electronic switching system (as shown in Fig. 19.3).

285

Three million PAM samples per second is about the upper limit of this technology at the time of writing. The 240 time slots handle up to 820 speech lines. In order to handle more than this amount, a space-division switching stage using reed relays must be connected ahead of the subgroup buses.

CONNECTION WITH PCM TRUNKS In the preceding illustrations, the "samples" that were switched were all PAM samples.

PAM samples are not used for transmission purposes because, being susceptible to noise and distortion, they do not have the advantages of pulse code modulated samples (PCM). PCM samples could, however, equally well be used in time-division switch. There is a prospect of considerable saving when both the exchanges and the transmission lines operate in a PCM time-division fashion. The network becomes so integrated that it is difficult to draw a sharp boundary between switching and transmission.

An economically attractive configuration for future telephone networks is to have subscribers connected to time-division concentrators that are linked to the nearest exchange by PCM (Fig. 18.5) lines. These lines would probably carry 24 channels like the Bell T1 carrier. The exchange would use time-division switching but would switch the PCM "samples" rather than the PAM. Queuing calculations for such a system, based on today's telephone traffic, suggest that 150 subscribers making calls at random could be connected to one such concentrator and a good grade of service would be given. That is, there would be a very low probability of a subscriber obtaining a "busy" signal because the concentrator and PCM line did not have a free channel. A smaller number of subscribers than this might be connected because they do not make calls at random. Nevertheless such a scheme would give much higher utilization of the wire pairs into the exchange than today's methods.

Any such arrangement would have to be linked into today's telephone network. Most of the present telephone exchanges are here to stay, for a long time (unfortunately). However, new capacity has to be added to most of the exchanges and local networks, and one way to do so is to operate a PCM system alongside existing plant. With traffic more than doubling every ten years and with most of today's plant being near saturation, we are likely to have a new network superimposed on the old.

Figure 18.6 gives an illustration of how this operation might be handled. The new units are small in size. The PCM exchange will normally

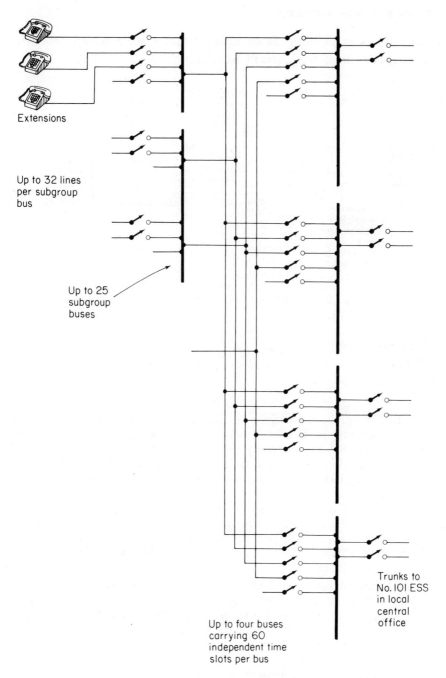

Extensions

Up to 32 lines
per subgroup
bus

Up to 25
subgroup
buses

Up to four buses
carrying 60
independent time
slots per bus

Trunks to
No. IOI ESS
in local
central
office

Figure 18.4. A time-division switching unit organization
like that in Fig. 18.3 but capable of handling 800 telephone
extensions.

287

fit into the same building as an existing exchange. Existing subscriber lines could be connected to the PCM concentrators.

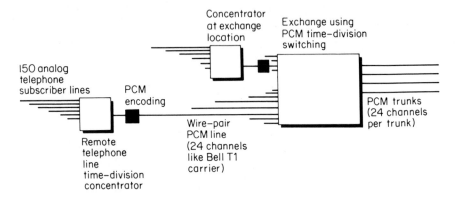

Figure 18.5.   Voice concentrators using PCM lines like the T1 carrier, and an exchange using PCM time-division switching.

ADVANTAGES OF TIME-DIVISION SWITCHING

The advantages of time-division switching can be listed as follows:

1. Time-division switching fits in well with time-division-multiplexed transmission. In a TDM network, it is likely to be lower in cost than space-division switching. Its cost will be lower when large-scale integration circuits are fully developed. However, when switching analog circuits, the cost of the required filters will keep the overall cost up.

   Unfortunately, the switching plant is under different management to the transmission plant in some common carriers. This practice made sense in the past, for the two were entirely separate, but now it may have tended to impede an integrated design for switching and transmission.

2. Time-division switching can be very fast, which is a desirable attribute for future data networks (see Chapter 8). Space-division switching can now also be made very fast if solid state switching is used (Chapter 19). Solid state space-division switching may also drop in cost greatly with the development of large-scale integration. However, space-division switching will always need more switch points than time division.

3. Time-division switching permits the construction of small-size and economic concentrators, plus switching units that can be employed in the design of private branch exchange and "centrex-type" systems.

4. A PCM switching stage can handle network signaling and control data as easily as speech. This fact permits designs in which some of the switching stages can be remote. There is much more scope for control of the switched network than with today's electromechanical exchanges.
5. As with PCM transmission, *data* can be handled naturally and quickly by PCM switch units.

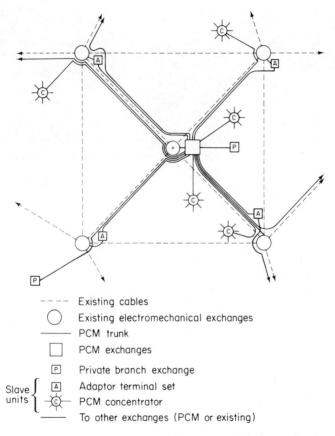

– – – –	Existing cables
○	Existing electromechanical exchanges
———	PCM trunk
□	PCM exchanges
P	Private branch exchange
A	Adaptor terminal set
C	PCM concentrator
———	To other exchanges (PCM or existing)

Slave units { A, C }

Figure 18.6. A new PCM network with PCM exchanges and concentrators, superimposed upon today's existing network. *Reproduced with permission from "Techniques of Pulse Code Modulation in Communication Networks" by G. C. Hartley, P. Morret, F. Ralph, and D. J. Tarran. Cambridge University Press, Cambridge, 1967.*

MARCONIGRAM

WORLD WIDE WIRELESS

Number 1

Feb 29th Midnight

Number of Words 95

MARCONI TELEGRAPH — CABLE CO. Inc.
IN CONNECTION WITH
MARCONI WIRELESS TELEGRAPH COMPANY
OF AMERICA

Send the following message "VIA MARCONI", subject to the terms on back hereof, which are hereby agreed to.

NEW YORK, MARCH 1st, 1920

GODFREY C. ISAACS,

MANAGING DIRECTOR,

MARCONI'S WIRELESS TELEGRAPH COMPANY LIMITED

LONDON.

MAY THIS FIRST MESSAGE WHICH OPENS COMMERCIAL WIRELESS SERVICE
BETWEEN AMERICA AND ENGLAND MARK AN EPOCH IN HISTORY FROM WHICH
THE ACHIEVEMENTS OF THE FUTURE SHALL DATE. COMMUNICATION IS
THE LEVERAGE WHICH SHALL LIFT THE WORLD TO BETTER UNDERSTANDING
AND THUS LEAD TO CLOSER TIES OF FRIENDSHIP BETWEEN ALL NATIONS.
IT IS THE MISSION OF OUR RESPECTIVE COMPANIES TO SO STRENGTHEN
AND IMPROVE THE WIRELESS SERVICE THAT DISTANCE SHALL BE MADE
NEGLIGIBLE AND COMMUNICATION PRACTICALLY INSTANTANEOUS.

BE 1   12.01 am
Wd

Edward J Nally.
PRESIDENT, RADIO CORPORATION OF
AMERICA

A copy of the first commercial transatlantic wireless message, 1920. Courtesy Radio Corporation of America.

19   COMPUTERIZED SWITCHING

The equipment that switches transmissions in networks of the future will be largely controlled by computers. In some cases, the switch paths will be through solid state circuitry so that there will be no moving parts. The advantages of computer-controlled switching are that it can be extremely fast—switching will take place in milliseconds rather than seconds as in today's telephone plant—and the computer can carry out many functions that would not be practicable before computing.

Present-day electromechanical switchgear has grown immensely complex. Some would say too complex because it accounts for many of the troubles the telephone companies are having at the time of writing. When one studies the immense complexity of a large electromechanical switching office, it seems surprising that it is possible to maintain it at all. Imagine what a modern computer would be like if you tried to build it of relays like the old accounting machines.

Clearly, computer-controlled operation can bring a great improvement. In one sense, it can bring simplification. Major strides are made in engineering when a new technique, although highly intricate internally, is used to bring sweeping simplification. The great engineer is the one who achieves simplicity and this fact is truer than ever in today's world of engulfing complexity.

There are economies of scale in switching as there are in transmission, although they are not as great. The cost per trunk of a 16,000 trunk switch is about half that of a 3000 trunk switch, with space-division switching. Time-division switching can bring greater economies of scale than space division. One key to switching economy, however, lies more in the elimination of operator assistance. In the United States, switching accounts for about 28% of the total investment in the telephone system plant. However, operator salaries cause switching to account for 45% of the cost of an average telephone call and 54% of a long-distance call [1]. Computers have helped and can help further in eliminating the need for operators.

The first major use of electronic switching came with the Bell System's No. 1 ESS (Electronic Switching System). This system is designed to replace, at an economic cost, central offices of a wide range of different sizes. It can handle from 10,000 to about 70,000 lines, with a maximum capacity of 100,000 calls in the busy hour [2]. For remoter areas with smaller numbers of subscribers, No. 2 ESS handles from 1000 to 10,000 lines. These systems are designed for mass production. At the time of writing, about 100 ESS 1 offices are in operation, and by the mid-1970s they will be installed at the rate of about 250 per year. Electronic exchanges not under computer control have also been designed for smaller exchanges. The TXE2 System installed by the British Post Office is designed for an exchange with only 200 to 2000 lines; the TXE3 System handles small exchanges with more than 2000 lines [3].

Other manufacturers have also been working on computerized switching. As data transmission on dial-up lines becomes increasingly common, exchanges controlled by computers—switching both data and voice lines —will become part of many communication networks. The private branch exchange as well as the public offices will become electronic.

ADVANTAGES The economic objectives of the No. 1 ESS are achieved merely in the replacement of existing central offices and an increased capacity in existing buildings. This advance in switching technology comes at an appropriate time, for many of the early central offices need replacement. No. 1 ESS needs a fraction of the floor space of its electromechanical equivalent; therefore, in some cases, the large expansion in switching capacity that is needed today can take place on existing premises.

In addition to these economic benefits, the flexibility of switching under program control has many advantages. Alterations in the exchange can be made by a change in memory contents of the computers and thus can be made very quickly in many cases. The manufacturing of the equipment is capable of more standardization because many of the variables that exist on electromechanical equipment are now variables in the program rather than in the hardware. An example of the power of program control was seen when a "centrex" service was added to the system. This service was incorporated by an addition to the programming. Growth of the exchange is made easier, for every No. 1 ESS can handle up to 70,000 lines. New subscriber numbers can be added quickly. In addition, the accounting procedures and traffic measuring procedures can be fully automated. Automatic features can also aid considerably in maintenance.

The computer program can make possible a variety of new features not on today's electromechanical exchanges—features that are attractive to the subscribers. It would be possible to set up conference calls with several parties joining in, by dialing. One could dial an extension in one's own home if it were equipped with more than one telephone. Bell System's first experimental electronic central office at Morris, Illinois, gave some subscribers "abbreviated dialing," which enabled them to dial certain selected seven-digit numbers using a two-digit code. This move proved very popular. Also, various methods were provided to allow the subscriber to have his calls automatically directed to another telephone. The subscriber could, for example, if he went to a friend's house for dinner, dial a special code and then his friend's telephone number. Thereafter all calls could be directed to this number automatically. When he returned home, he would cancel the rerouting by dialing another code. Alternatively, he could inform the telephone company of the times and numbers for rerouting.

In addition to some of these facilities, the *System AKE* stored program exchange manufactured by L. M. Ericcson in Sweden has an automatic call transfer. Incoming calls may be transferred to another specified number, either when the subscriber is busy or when he does not answer. Again, if you dial a subscriber on this system and he is busy, you may then dial a code that instructs the computer to dial you as soon as he is free, and then automatically redial him. The need for redialing is dispensed with. System AKE also has an alarm clock service. You dial a four-digit time at which you wish to be woken by the computer. Subscribers may dial directly to private automatic branch exchange extension telephones without operator assistance. System AKE can be set to gather a variety of statistical information about subscribers' calling habits. The system can list particulars of incoming or outgoing calls on particular numbers. It has a facility for recording malicious calls. If a subscriber receives a malicious call, he does not replace his handset for a time and system then records his number and the number of the calling party.

Other new facilities are planned. Such is the flexibility of operating under programmed control.

BASIC
PRINCIPLES
The essential elements of a computerized switching office are shown in Fig. 19.1. The incoming lines all enter some form of switching matrix [1], which could be built from crossbar switches, reed relays, or solid state switches with no moving parts. The switch controls for setting up different paths through the matrix are operated by the computer [2].

There is a mechanism for constantly scanning the activity on the lines [3] and informing the computer program of significant events. The program must be informed when you pick up your telephone handset; it must read the number you dial and must be informed when you replace your receiver after a call. Similarly, it must detect and interpret the signals from other switching offices that arrive on the trunks.

There are various signals that it must send down the lines, such as "busy" signals ("engaged" signals in British parlance), signals to make telephones ring, dialed numbers, and on-hook/off-hook signals, which it sends to other switching offices. The signaling equipment [4] might be regarded as one of the computer output units in this type of system.

Information about calls must be gathered for billing purposes, and details about the uses of the network will be filed for statistical analysis. These will be recorded by the computer [5].

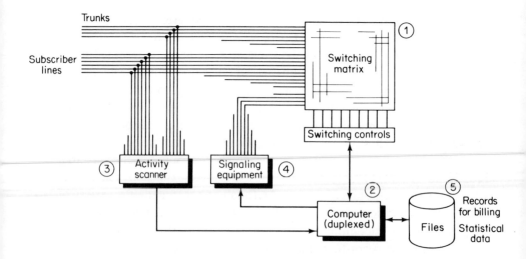

Figure 19.1. Basic elements of a computerized switching system.

ORGANIZATION
OF ESS 1

Figure 19.2 shows how these principles are applied to the Bell System's No. 1 ESS.

The subscriber lines (1) and trunks (3) enter a switching network (2) in which the interconnection paths between lines or between lines and trunks can be set up. The switches used are reed switches, called *ferreeds*. Magnetic contacts encapsulated in glass, free from dust and corrosion, are opened and closed by windings outside the glass. They are arranged in pluggable units containing an 8 × 8 array of switches.

The switches for making the appropriate line connections are opened and closed under control of the program in the computers (7).

The switching network is organized into frames, and each frame has its own *switching controller* (5). The controller sets up the appropriate line paths in the frame it controls. Both the switching controllers and the computer are duplicated for reliability reasons, as is much of the remainder of the system. When one such component fails, its counterpart takes over.

The computers have two types of memory unit. *Semipermanent memory units* (9) contain the programs and fixed data about the lines. This is "read-only" during normal operations and so cannot be accidentally overwritten by a program error or hardware failure. *Temporary memory units* (8) contain information about the calls in process and other information that is changing during normal operation. Both types of memory unit, and their contents, are duplicated.

The computer knows the status of the lines, trunks, and signal receivers by means of a continuing scanning operation, carried out by the *line scanners* (4), which are also duplicated. After each scan, the computer writes information about the line status in the temporary store.

Each line and trunk is equipped with a sensing device called a *ferrod*, (13), which can detect telephone on-hook and off-hook conditions. The ferrod is a current sensing device consisting, in essence, of a ferrite rod with two identically wound solenoid coils through which the line current flows. Two other loops of wire are threaded through two holes in the ferrite rod; these are used for sensing whether or not line current is flowing. The scanner sends an *interrogate pulse* down one loop and detects whether there is a corresponding *read-out pulse* down the other loop. When no dc current is flowing in the line, there will be a strong read-out pulse, but when current is flowing, the pulse will be suppressed. The scanners, then, scan these ferrods, and the information they obtain about line status is read into the temporary storage by the computer.

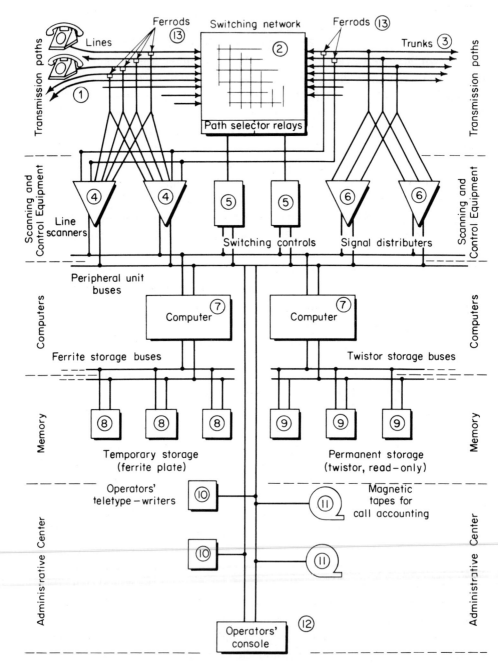

Figure 19.2.   Basic configuration of No. 1 ESS.

The scanners are provided in modules that scan 1024 lines. A scan takes place every 100 milliseconds, in which each subscriber line is examined to detect call originations. If an origination is detected, that line is connected to a device which interprets the dialing signal, either dial pulses or Touchtone. During this period, a line is sampled every 10 milliseconds. Once the call is connected, the line is again scanned every 100 milliseconds, the purpose now being to detect when the call is terminated. When the caller hangs up, the line is disconnected by the computer.

As the scanning takes place, the results from the ferrod sensors are sent to the computer, 16 at a time, and compared with the results of the last scan, which are stored in temporary storage. If there is a change, then the computer branches to the appropriate program to handle it.

It is necessary to send a variety of signals on trunk circuits. No. 1 ESS uses a *signal distributor* (6), again duplicated, for this purpose. Under computer program control, this distributor connects the appropriate signals to the trunks as required.

At the bottom of Fig. 19.2 are the devices used by the operating personnel. These devices include alarms, line test facilities, displays, and control circuitry for maintenance; a teletypewriter (10), duplicated, allows the operating personnel to communicate with the computer. A magnetic tape unit (11), duplicated, writes on tape all data related to billing customer calls. This information is then processed on an off-line computer. Finally, there is a device for changing the information in the semipermanent storage.

THE COMPUTERS   The engineering of the computers in No. 1 ESS is considerably different from that of machines familiar in today's data processing world. The major factor leading to this difference is the need for extreme system reliability. A design objective was that the number of periods of overall system failure should not exceed 2 hours over its 40-year life [5]. Furthermore, the system should be left unattended for long periods of time. These criteria are far beyond the standards of today's conventional data processing.

Although system outage is regarded as catastrophic, individual errors are not. Individual errors in conventional data processing may be very serious, but in an exchange they are merely a nuisance. The subscriber may redial his call. One does not want such errors to occur often, but certainly subscribers do have to redial calls occasionally on today's electromechanical plant. Most subscribers, when their dialing fails to reach its destination, blame themselves rather than the machinery the first time it happens—thinking they must have misdialed. This facet of dialer psychology is useful in planning the error and recovery procedures in a computerized exchange.

The computer components have therefore been chosen with breakdown-free operation as the main criteria rather than speed or other factors. All system components other than those concerned with a single line are duplicated, including the data buses (shown in Fig. 19.2). The computer is organized and its components designed so that faults can be found and corrected as quickly as possible. When a fault develops, it must be corrected rapidly to minimize the probability of a duplex failure. During normal operation one computer is "on line," carrying out the switching work; the other is in active standby. Both the standby and the on-line machine are automatically monitored continuously so that any fault is signaled as soon as it occurs. When a fault occurs in any of the duplicated components of the system, a switchover occurs to the alternate unit automatically so that the telephone actions are interrupted for only a short duration.

The instruction word length is 44 bits, but 7 of these are used for automatic error detection and correction. The circuitry automatically corrects single-bit errors and detects double-bit errors. In addition to this checking, the system contains special circuitry that compares the execution of instructions in the two computers. If a difference is detected, the engineers are notified. Also, all transfers to and from the scanners and other peripheral devices are checked.

The circuitry of the computer is designed very conservatively to give the maximum reliability. It is considerably slower than conventional data processing machines. The cycle time of the computer is 5.5 microseconds. The storage units are of unusual design, again planned to maximize reliability. The majority of the storage is of a read-only nature and program malfunctions cannot damage it.

To handle the peak-hour traffic of a busy office with 65,000 lines, it was estimated that the system could not spend more than 5000 machine cycles per telephone call [6]. In order to meet this requirement, it was necessary to design an instruction set with powerful instructions that carried out several functions simultaneously. In addition to general-purpose data processing instructions needed to make the machine flexible enough to handle a variety of possible future demands, it also contains special instructions for reading the scanners and performing operations on their information, as well as for sending output to the switching control equipment and the signal generators. The machine thus has a mix of general-purpose and special-purpose instructions.

Some input-output programs have to be executed at strictly determined times—for example, when a subscriber is dialing, the dial pulses are sampled every 10 milliseconds. This action is governed by a clock that interrupts the program being executed, so that control is transferred to the appropriate priority program. When the condition that caused the interrupt has been dealt with, control is returned to the program that was interrupted.

As on more conventional computers, a variety of conditions can cause an interrupt, including the detection of an error by the checking circuitry. The computers have three levels of priority for normal error-free processing. Interrupts can normally occur only at 5-millisecond clock intervals, and snatch control away from lower level programs, giving control to a higher-priority program. When the higher-priority work is completed, the computer returns to the program that was interrupted. There are another seven levels of priority, however, for error conditions. These seven levels of interrupt can occur at any instant when a fault is detected. They trigger a variety of procedures designed to give automatic switchover to a fault-free configuration. Duplicate units can be switched in times comparable to the cycle time of the unit. Thus for most of the failures that occur, there will be no break in the processing of the majority of the calls. An abnormally large proportion of the programming in this system is concerned with reliability—more than half of all the instructions.

The programming proved to be a major problem in the installation of ESS 1. The number of instructions grew much larger than anticipated, and this fact gave rise to difficulties in handling the intended volume. There were delays and schedule slippages. For a time, some of the installed systems had troublesome bugs in them. But all of this is hardly surprising to anyone who has lived through the installation of equally complex real-time systems.

ESS No. 2   ESS No. 2, designed to handle 1000 to 10,000 lines, is basically similar [7]. Its intended use is in nonmetropolitan areas, where it will stand largely unattended. For this reason, maintenance tests, traffic and plant measurement, translation changes, and other factors were designed so that they could be handled remotely. The engineers only visit the office when repairs, replacements, and reconnections are needed, and for occasional preventative maintenance.

Economies are gained in ESS No. 2 by not giving it a high-call handling capacity. In general, the processing has been kept as simple and economical as possible. Some features are needed in a rural area that are usually not found in a metropolitan area—for example, the ability to handle party lines with many parties.

THE PROCESSING
OF A TYPICAL CALL

Let us now describe the processing of a typical call in No. 1 or No. 2 ESS to show how computerized exchanges work.

1. When a subscriber picks up his telephone, the electronic exchange must detect that service is needed. The line in question will be inspected by the scanner every 100 milliseconds, using ferrod sensor. When the customer lifts his telephone, the flow of current is detected.

   The computer, comparing the 24-bit word it receives from the scanner with that it received 100 milliseconds ago and stored in the ferrite store, will detect a change. The computer will then examine the change more closely and will see that one bit has changed, relating to the line in question.

2. The computer next connects this line to a digit receiver by instructing the switching controls to set up an interconnection path. The ferrod sensor is disconnected from the line until the completion of the call, and a dial tone is connected to the line.

3. On hearing the dial tone, the customer will dial. If he has a Touch-Tone telephone, this type of digit receiver will be used; otherwise he will use a rotary dial pulse receiver. The computer will read the digit receiver for the line every 10 milliseconds; the clock interrupts every 5 milliseconds, thus ensuring that none of these readings is missed.

4. If the telephone number dialed is to a telephone connected to this central office, the computer will look at the condition of that line, as recorded on the last scan. If it is busy, the computer will connect a "busy" tone to the calling line. If not, it will switch a ringing tone to the called line, again by instructing the switching controls to make an interconnection to a ringing circuit.

5. At this time, both the calling and the called line must be supervised. If the caller hangs up, this act must be detected and the call abandoned. Such detection is not now done by the ferrods and scanner but by other circuits associated with the digit receiver and ringing-tone circuit. Signals must be supervised on the trunk to detect successful connection, or a busy condition.

6. When the called party answers, the ringing tone is removed from the line and the computer instructs the switching controls to complete the necessary interconnection.

7. While the parties talk, the calling and called lines are supervised so

that the termination of the call can be detected. On a trunk call, this step would be done by the ferrods on the trunk circuit.

8. When the call is terminated, the computer disconnects the circuit and restores the lines to their former status. The 100-millisecond scanning takes place as initially.

9. The computer then completes a record for accounting purposes. To do so, it will have timed the trunk call.

In all cases throughout this operation, signals on the lines and changes in line state do not of themselves cause system actions to take place, as was the case on electromechanical exchanges. Such conditions are detected by the computer, by scanning, and the computer decides what to do. In this mode of operating lies the potential for building a more "intelligent" communication network in the future, capable of many operations not possible with today's telephone exchanges.

PRIVATE BRANCH EXCHANGES

As we shall see, there is much to be gained from computerizing private branch exchanges as well as the public offices. Here, however, we run into economic problems. Systems like No. 1 and No. 2 ESS are economical only in large sizes. No. 2 can be justified when it can handle more than a thousand lines. About 90% of private branch exchanges have less than 200 extensions. How do you justifiably computerize these?

There have been two types of approaches. Both give important pointers to the directions in which this technology is likely to progress in the future. The first is a *time-sharing* approach. The computer is still large, like ESS 1, but takes over the operations of what would have been many private branch exchanges. The switching is controlled in the local central office rather than on the subscribers' premises. The Bell System No. 101 ESS operates in this way. Using technology similar to No. 1 ESS, it controls up to 32 time-division switching units in separate subscriber locations.

The second approach is to make the exchange an adjunct of a data processing system, as is the case with IBM's 2750 Voice and Data Switching System, which is marketed in Europe but, unfortunately, not in the United States at the time of writing. This system is a fully electronic private branch exchange that, in addition to interconnecting telephones, connects terminals, data collection equipment, and contact-monitoring points to a data processing system.

ESS No. 101
Figure 19.3 shows the connection of sub-scribers to a No. 101 Electronic Switching System at the central office location [8]. The system can handle up to 32 separate subscriber locations. Most subscribers have a unit that permits the connection of up to 200 telephone extensions. Other larger units can handle, 340, 820, or 3000 extensions.

The switching is done in the small cabinet shown in Fig. 19.4. This uses time-division switching and is controlled by the remote computer. The organization of the switch unit was shown in Fig. 19.3. Fifty time-division paths through the switch unit are available to the 200 telephone users, a ratio that is usually more than adequate. If the average subscriber uses his telephone one tenth of the time, then the probability of failing to obtain a circuit will be extremely low.

The digits dialed or keyed by a telephone user go to the ESS 101 computer on one of two data lines connecting the computer and the switch unit. The other data line is used by the computer to control the switching —in other words, to assign the time slots to the appropriate lines. In this way, the switching is under control of the computer program. There are voice trunks from the switch unit to the computer, for incoming and outgoing calls, and these trunks are switched in the same way.

SOLID STATE SWITCHING
Although ESS 101 is particularly interesting because of its time-shared computer control of remote TDM switches, the IBM 2750 Voice and Data Switching System is interesting because of the technology of its switching matrix. It is entirely solid state. It uses space-division circuit switching, but there are no moving parts and hence the switching time is extremely fast.

The crosspoints of the switch matrix are built from PNPN thyristors, as shown in Fig. 19.5. When a pulse is sent to the gate lead, the crosspoint is switched ON and current may flow through the anode-cathode path. The crosspoint will remain ON as long as current flows in this path. The components in Fig. 19.5 are on one silicon chip, and many such chips provide a switching matrix on one circuit card, as in Fig. 19.6.

Because this is space-division rather than time-division switching, a physical circuit path is set up and any type of signal may flow through it. The circuit may carry data at a very high bit rate. Such is not the case when time-division switching is used, as in the ESS 101 switch units with a sampling rate of 12,500 samples per second. As discussed in Chapter 8,

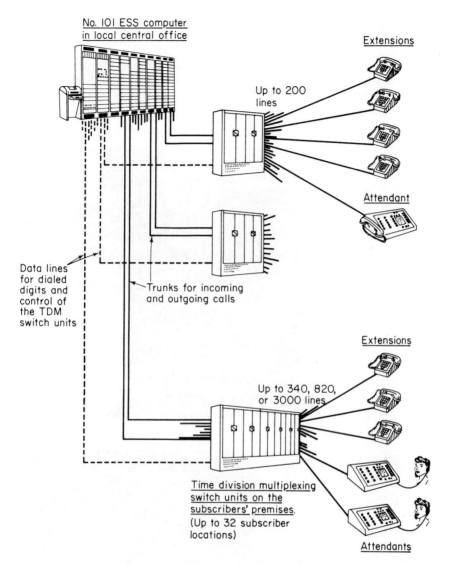

No. 101 ESS computer
in local central office

Extensions

Up to 200
lines

Attendant

Data lines
for dialed
digits and
control of
the TDM
switch units

Trunks for incoming
and outgoing calls

Extensions

Up to 340, 820,
or 3000 lines

Time division multiplexing
switch units on the
subscribers' premises.
(Up to 32 subscriber
locations)

Attendants

Figure 19.3. The connection of subscribers to a No. 101 ESS computer that operates in a time-shared fashion to provide the functions of up to 32 private branch exchanges.

very fast switching is desirable for some types of data networks. Solid state switching of this type can provide it.

THE IBM 2750 SYSTEM   The IBM 2750 System is designed to carry a wide variety of transmission types. It is intended for major data processing users, such as banks, insurance companies, manufacturers, and large laboratories. Sev-

303

Figure 19.4. The time-division switching unit on the subscriber's premises when an AT&T ESS 101 is used to replace private branch exchanges. This is the unit shown connected to 200 telephone handsets in Fig. 19.3. *Courtesy of Bell Telephone Laboratories.*

Inside view.

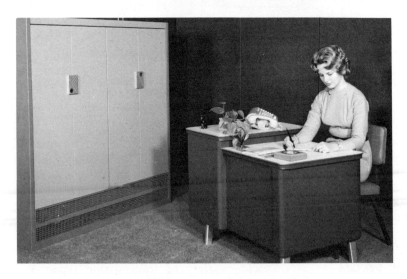

Outside view with attendant's console.

eral such organizations are using the system in Europe. It provides them with a private branch exchange in which data and voice capabilities are integrated. It can serve over 700 extension lines and more than 80 outgoing trunks.

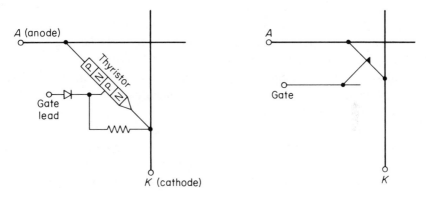

Figure 19.5. Crosspoint of a solid-state switching network that was built using thyristors.

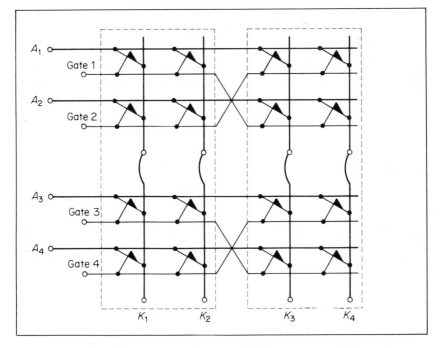

Figure 19.6. The thyristor crosspoints are assembled on matrix cards such as this 4 x 4 card or on larger cards.

Any terminals or data processing machines may be interconnected by dialing from a telephone, the telephone being free for voice transmission once the connection is made. For inexpensive data collection, pushbutton telephones may be used, and the 2750 assembles the messages, checks each character, adds control information, such as date, time, and terminal identification, and directs the data to a computer, tape punch, or other device. Voice answerback may be used for controlling data collection, or other terminal operations. The machine can read the setting of switch contacts, on the customer's premises or remotely connected. It can similarly operate contacts. Such contacts may be used for all types of different purposes—for example, fire or burglar detection, process control functions, or remote meter reading.

The 2750 can be equipped with a paging unit, which consists of a radio transmitter and up to 90 pocket receivers. Any extension user can page a person directly by dialing. That person hears "beep" signals from his pocket device, and when he responds, he is automatically connected to the caller.

A record voice announcement attachment is available; it could respond to external callers in certain circumstances or could send messages to extension users under computer control.

For reliability, two computers are used in the system. At any one moment one computer is "active" and the other in "standby." The network is under control of the active computer, and the standby computer is always ready to take over. The standby computer continuously checks its own circuitry and monitors the active computer. If the active computer fails, an automatic switchover takes place and the standby machine becomes "active."

PICTUREPHONE SWITCHING As was discussed in Chapter 4, AT & T intends to offer its Picturephone subscribers the same switching facilities as its voice subscribers, and this process must be handled within the framework of the existing networks. The ESS computers will be used for Picturephone as for voice.

When ESS No. 1 is used, the arrangement will be similar to that in Fig. 4.3. The four video wires will be switched to the voice wires. The voice wires will be scanned as described earlier for the on-hook/off-hook and dialing signals, and when the computer switches the voice lines, it will also switch the video lines in parallel.

Figure 19.7 shows the arrangement when No. 101 ESS is used. The time-division switch units of Fig. 19.3 cannot be used for video because the sampling time is not sufficiently frequent. Therefore a separate video

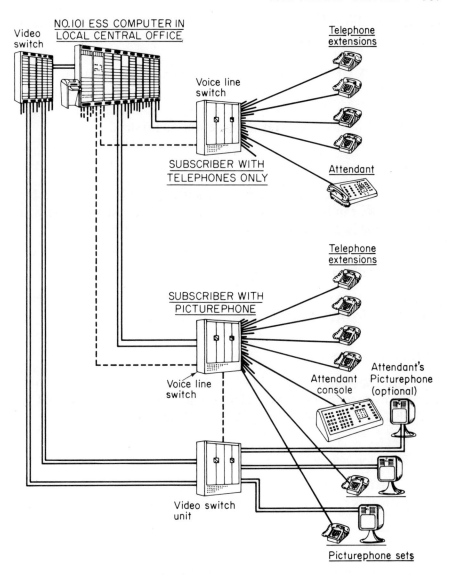

Figure 19.7. The addition of Picturephone switching to subscribers using a No. 101 ESS.

switch unit is used alongside the voice switching unit. The former does not use time division as does the voice switching unit but conventional four-wire reed relays. These relays are under the control of the ESS 101 computer, and the same data link is used as for voice. The switching again takes place in parallel with the voice switching.

In either arrangement a very high speed data terminal could equally be switched through the Picturephone network, using the telephone for establishing the connection.

ADVANTAGES OF COMPUTER CONTROL Following are some of the functions that become possible when a private branch exchange is under computer control. All the functions on this list are implemented today on some working system.

1. Conference calls

Calls in which parties on three telephones (or Picturephones) intercommunicate can be set up from an extension. The systems could be designed so that larger numbers of users are able to confer.

2. Automatic rerouting of calls

When a person leaves his office, he can instruct the computer to reroute calls to a different extension.

3. Calling waiting

A customer whose line is busy is notified of an incoming call to his number by a series of short "beep" sounds. He can be connected temporarily to the new call and can queue it if he wishes.

4. Hold for inquiry

An extension user can hold an external connection while he dials another number—for example, to make an inquiry.

5. Call transfer

An extension user can automatically transfer a caller to another extension. If this extension is busy, a warning signal (short beeps) can be sent to inform the user that there is a call for him.

6. Abbreviated dialing

A frequently dialed local or long-distance number can be called by dialing two digits only. The computer translates the two digits into the full number by looking up a table for that extension. This practice is particularly valuable for long-distance numbers which would need 10 digits dialing, or 11 (first digit 9 to access public network).

7. *Multiple classes of service*

Restrictions may be placed on what calls an extension may receive or initiate—for example, a certain extension may receive only internal calls. Certain extensions may be prevented from making outgoing calls or may only make local calls, not costly long-distance ones.

8. *Extension hunting*

This process enables a group of extensions to be associated so that if any one is dialed and is busy, then the others are automatically scanned until a free one is found. Each extension in the group can also be dialed individually. This feature is useful when, for example, different persons are equally qualified to answer certain calls.

9. *Priority access to operators*

Some extensions can be given priority over others when calling the operators. This feature would be used to give rapid and perhaps more competent operator service to management.

10. *Alarm clock functions*

A telephone user can request a call at a future time which he keys in.

11. *The ability to change configurations*

Because the network is under program control, extension number and other aspects of the call routing can be easily changed from a console. No rewiring is needed when you move to another office.

FUTURE NEEDS IN SWITCHING SYSTEMS A variety of new needs are likely to arise in telecommunication switching. Some of the possible ones are as follows.

Further facilities could be provided for telephone (and other) subscribers in the computer-controlled offices. Good use could be made of a feature in which the switching office *originates* calls. For example, I can waste much time attempting to call a man whose telephone is constantly busy or unattended. Sometimes I can spend an afternoon and fail to reach him, and people tell me that they sometimes fail to reach me. It would be useful if I could instruct the computerized exchange (private branch exchange or central office) to interconnect us as soon as his telephone becomes free, and perhaps to dial an apartment at intervals so that I catch somebody as soon as he comes home.

Again, a telephone-answering service is a useful service for busy or highly mobile people, but few people have one. This is a service that could be provided by a central office at lower cost than installing one's own equipment. Messages for subscribers would be recorded in computer storage and relayed to them when they dial in. The computer might be instructed to call them at a given time if there are any messages for them.

The same problems arise with data transmission, and as discussed in Chapter 10, data storage will be a useful facility to have in switching offices of the future.

When the Picturephone service is in full operation, subscribers may find conference calls particularly valuable. They will be able to have "meetings" without traveling. This operation poses switching problems. There may be a meeting room in the future with many wall screens, but it is probable that most of the participants will have only one screen. One can hear everybody at the same time, and no sound switching is needed once the conference connection is set up. However, if only a single screen is used, one cannot see everybody. Whose face does one have on one's screen and what switches it there? Possibly a certain member of the Picturephone meeting will control the switching. Another intriguing possibility, however, is that discussed in Chapter 4 in which the switching would be voice-activated. The screens of the conferees will automatically show the face of the person who is speaking, like automatic movie editing.

A big question for the 1970s is "To what extent will data switching and voice switching be integrated?" It is interesting to see the way Picturephone switching has been grafted onto existing voice-switching systems. This technology could provide public switching for very wideband data channels, at least in the United States and Canada, where Picturephone is being marketed. If other countries, as some of them state, will not have video-telephone facilities until 1990, then some other provision for very high speed data switching is needed.

A major concern in data switching is for real-time systems; for these, as mentioned in Chapter 8, a fast switching time is desirable. The telephone switching speeds of today are too slow. If all offices were under computer control and were designed to operate rapidly enough for real-time needs, there would be no problem. However, it will be a long time before that is the case. This fact, if no other, might prevent the integration of public data and telephone networks. It is possible that in some countries the same trunks and subscriber lines might be used but with separate switching facilities.

It is appealing to provide a network with special facilities for data terminal users at the switching offices. Particularly attractive is the idea that the facilities should permit terminals of different types to be interconnected. The "interface computer" of D. W. Davies (reference 2 in Chapter 10), which would "talk to each subscriber terminal in its own language," has great potential even if the rest of his message-switching scheme does not come into existence. This idea will become increasingly attractive as the cost of logic circuitry and computer storage drops (Figs. 16.4 and 16.7).

When terminals for the general public finally become a mass market, there will be a great proliferation of different computer systems that the subscriber can dial. Frequently he will be in need of elaborate directory assistance. No doubt a "Yellow Pages" of computer services will be available, but perhaps the best source of help will eventually be yet another computer and its data bank. Subscriber terminal conversations may begin with the common carrier's own computer, which connects them to the machine they need and which often gives them real-time directory assistance in finding it!

REFERENCES

1. President's Task Force on Communications Policy, Staff Paper No. 1, Washington, D.C., June 1969.
2. W. Keister, R. W. Ketchledge, and H. E. Vaughan, "No. 1 ESS: System Organization and Objectives," *Bell System Tech. J.*, September 1964.
3. "Electronic Switching Systems. 1. TXE2 Electronic Exchanges," a booklet published by the GPO, London.
4. "Stored Program Controlled Switching, System AKE," manufacturer's booklet from L. M. Ericsson Telephone Company, Stockholm 32, Sweden.
5. R. W. Downing, J. S. Novak, and L. S. Tuomenoksa, "No. 1 ESS Maintenance Plan," *Bell System Tech. J.*, September 1964.
6. J. A. Harr, F. F. Taylor, and W. Ulrich, "Organization of No. 1 ESS Central Processor," *Bell System Tech. J.*, September 1964.
7. No. 2 ESS, a complete issue of the Bell System Technical Journal on this subject. October 1969.
8. W. H. C. Higgins, "A Survey of Bell System Progress in Electronic Switching," *Bell System Tech. J.*, July–August 1965.

SECTION **IV**

**THE MORE DISTANT
FUTURE**

The earliest form of mobile radio, 1901. Courtesy The Marconi Company, Ltd., England.

20   MOBILE RADIO
        TRANSCEIVERS

For many years science fiction has been full of small, portable, radio-operated devices like Dick Tracy wristwatch intercoms, and paging devices that find James Bond no matter what he is up to. With the transistor, portable radio receivers suddenly became a reality—often a menace. Now, with mass-produced LSI circuitry, more elaborate portable devices are possible. A miniaturized Touchtone telephone keyboard *could* be worn on the wrist. Once again, in the age of the computer, all manner of applications are possible for portable machines.

   In cars, particularly, it would be useful to have an inexpensive terminal and radiotelephone. The Touchtone keyboard could be used as in Chapter 11 for obtaining real-time traffic reports and avoiding areas of blockage or congestion. The driver returning home could switch on his apartment air conditioner, start the electronic oven, or talk to his wife or children. When his car breaks down, he could radio for help. He could be warned automatically of fog or accidents ahead. If he has an accident, a hospital could be telephoned and fast arrangements made. Transmission of information from his identity card would enable computers to look up his medical records for blood group, allergies, and so on. For salesmen or delivery vehicles, route scheduling could be carried out and orders keyed directly into a terminal in the vehicle. A typical organization can increase the utilization of its delivery fleet by 50% with real-time radio control. "Dial-a-bus" and other such computer-controlled schemes [1] become economically attractive for suburban transportation.

THE RADIO
SPECTRUM

There is one snag, however. The radio spectrum, managed as it is today, is overcrowded in areas that might be used for radiotelephone applications. In London, at the time of writing, the number of unfulfilled applications for mobile radiotelephones exceeds the total number in use. Various U.S. government reports [2, 3] have indicated that within metropolitan areas there is a critical impending shortage in frequencies allocated to certain two-way radio services, including police, taxicabs, fire engines, ambulances, and business users.

There are, however, as mentioned in Chapter 13, ways out of this dilemma. They lie in the following areas:

1. Changes in the way the radio spectrum is allocated, avoiding the allocation of large blocks to services that can be handled by means other than radio.
2. New designs of transmission systems for more efficient spectrum utilization. The use of strictly localized transmitters and a combination of radio with land-line transmission and switching.
3. Systems in which many users share several frequency paths, with dynamic allocation of channels carried out automatically.
4. Extending the use of the frequency spectrum into the field of millimeter waves, which will multiply the total available spectrum many times but will permit only very short distance radio transmission.

Appropriate action along these lines could increase the number of mobile transmitters by a factor of 100 or more in the not-too-distant future. New technology and near-complete penetration of cable TV, freeing much of the television frequency requirement, could bring a much greater increase.

In the United States today there are approximately 100 million vehicles and 100 million telephones. There are 6 million radio and television transmitters, of which about 2.5 million are mobile two-way radios.

The aim of land mobile radio systems is to provide a circuit between all users who need one. A short-term objective would be to permit telephone (and data) calls to be made to and from *all* vehicles. A longer-term objective may eventually provide this service between every person in the United States, so that everyone could carry a small personal transceiver (combined transmitter and receiver). For the former goal, the number of transceivers using the airwaves would have to be increased by a factor of about 50; for the latter goal, by about 100 [4]. As we shall see, these goals do not appear unreasonable; they could be reached within

20 years. Figure 20.1 shows the results of a study estimating the growth in numbers of radio transmitters in the United States [4]. It indicates that if the demand for mobile services increases as anticipated and the spectrum space is made available, there could be one transmitter per person by 1990. If large numbers of transmitters came into use, new types of functions would become available to the car or pocket transceiver. Those new functions would in turn increase the market demand. As with radio, television, and the telephone in their day, a snowballing effect would set in until most of the public owned one.

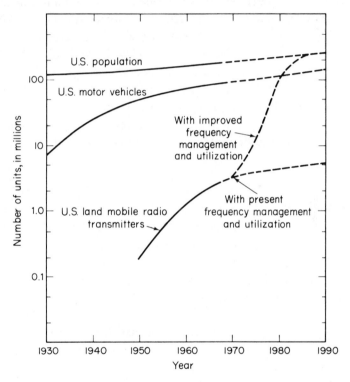

Figure 20.1. Projected growth in the number of land mobile radio transmitters in the United States. *Redrawn with permission from reference 4.*

LINKAGE TO THE TELEPHONE NETWORK

Each portable transceiver would be connected to the telephone network land lines at the closest possible point. If a sufficiently large number of land-based transmitters were in existence, the radio path would be fairly short. Once the call reached the land lines, it would travel over the Bell System or equivalent facilities, and possibly over satellites, like any other telephone call. The car transceiver would probably have a standard Touchtone telephone keyboard. The pocket or wrist device would have a miniaturized version of the same keyboard. Either could use the keyboard as a computer terminal, with overlays as described in Chapter 11. One can imagine carrying a wallet full of keyboard overlays the same size as credit cards.

Most mobile radio systems today use a "net," which operates over a limited area with a limited number of subscribers. Such a net would often use one frequency for all the subscribers. A manual control center is used, and discipline between the subscribers must be maintained. Such a net may be connected to the telephone network via its control point. Increasingly, however, individual subscribers will be connected on their own frequency. They may be connected on a dynamically allocated frequency group designed to give a specified "grade of service"—that is, probability of immediate connection. They could dial a conventional (nonmobile) telephone number, or could dial a mobile transceiver linked to a different land-based transmitter. In this way, nationwide dialing and eventually multicountry dialing, mainly via satellite, would be possible.

MOBILE TELEPHONE NUMBERS

A change in the telephone-numbering scheme would be needed if subscribers carry their transceiver from the zone covered by one land transmitter to that of another. In a scheme discussed by F. R. Eldridge [4], each private transceiver would be permanently associated with its owner. The number of the individual would be dialed, or perhaps even his name would be keyed into the call-originating transceiver—"John Smith 42" where there are many John Smiths. When his prime location is different from that of the call originator, it would be necessary to key also a number giving his prime location, just as an area code is dialed on today's telephone network.

It is envisaged that the switching on such a system would be done entirely by computers. These computers could be a development of the Bell System ESS 1 or they could be new machines, perhaps using solid state switching as in the IBM 2750 private branch exchange.

Each subscriber would be listed in the machine serving his prime location. When that person is "dialed," the call would be transmitted to the computer at his prime location and the computer would look up his whereabouts. It would have a record that tells whether he is within the area covered by that computer. If he is within that area, the record would say which transmitter zone he is in at present. The call would then be switched to this transmitter and the appropriate transmission frequency selected.

If the called person is not in the area covered by his prime-location computer, the record examined by that computer would say where he is, if that fact is known. The call would be routed by land line or satellite to the computer serving the area in question. This computer would then look up which transmitter zone he is in and switch an appropriate circuit path to that transmitter.

In this way a call dialed to a person could reach him at any location served by the system. If he is in his car, the radiotelephone could ring. If he is on the Metroliner train to Washington and is wearing a pocket transceiver, the call would come over the train's radiotelephone and be transmitted to its passengers over antenna wires within the train. When he is on an airliner to Hawaii, the call would go via aeronautical radio-telephone if one of the channels to that plane is free. If they are all busy, a recorded voice from the computer would tell the caller to try again in 5 minutes because his quarry is on an airplane. Alternatively, it could offer to call him back.

The whereabouts of the called person may not be known to the net-work. Also, he may have switched off his transceiver because he does not want to be disturbed. Then the stored-computer voice would inform the caller that such was the case.

KEEPING TRACK OF MOBILE TRANSCEIVERS

How does the computer network know the location of the called parties?

There must be a mechanism for keeping track of the portable transceivers. When they move from one transmitter zone to another, this fact must be recorded. Possibly the easiest way to do so is to design each transceiver so that it transmits its identification number at intervals. This number will be picked up at the nearest fixed transmitting station, which will add its own code and send the signal over land line to the area computer. The address transmission from the portable transceiver may be received by more than one fixed station. Each station may add a digit to the address to indicate the strength of the signal it received. The area computer would record which fixed transmitter received the best signal from each portable trans-ceiver. It would then switch circuits to that transceiver, using the best fixed transmitter.

When a subscriber moves from the zone of one transmitter to the zone of another, the subscriber record in the area computer would be appropriately modified. When he moves from the region covered by one area computer to that of another, the computer would transmit messages to the computer in his prime location, saying which area computer could now contact him.

The linkage between the fixed transmitters and the area computers may be over land line or cable, possibly over microwave paths. The links between the computers would be over the public telephone network, as with calls routed between toll offices today (Fig. 20.2). The area computers could, in fact, be located at the site of today's toll office. Probably the best way for a system like this to come into being would be as an outgrowth of today's telephone network, although it is perhaps likely that independent entrepreneurs will bring it about, as happened with Microwave Communication, Inc., the Datran System, and the facsimile mail companies.

THE RADIO LINKS

It is advantageous to have the radio links cover only a small area, so that the same frequencies can be reused in many different areas, as in Figs. 13.6 and 13.7. The fixed transmitters will be low in power and designed to cause as little interference as possible, unlike most of today's land mobile services and radiotelephone transmitters.

Each transmitter will be less expensive than today's fixed transmitters for mobile services, but there will have to be a large number of them. The transmitters will take a variety of different forms, for they will have to reach vehicles in city streets, vehicles on long-distance highways, and perhaps people with portable transceivers in buildings, home, on trains, and so on. Many different designs for low-power distributed systems are now being explored. Highly directional antennas are being designed to beam signals down city streets. The shielding properties of large buildings could be taken advantage of. Long twin-lead antennas, which could extend along highways or city streets, are being investigated. These antennas are being used for transmission in subways and tunnels—for example, by the New York City Transit Authority. Wire antennas are also being used within buildings for paging systems. Such antennas can confine their radiation to within a few hundred feet. Small walkie-talkie transmitters are being used to enable commentators to follow golf matches and other sporting events. Different means of signal distribution will meet different situations, and the radiation patterns must be carefully tailored to give efficient use of the available frequencies.

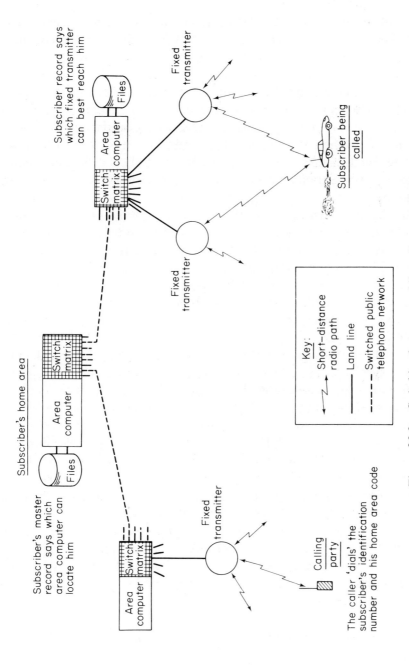

Subscriber's home area

Subscriber's master record says which area computer can locate him

Subscriber record says which fixed transmitter can best reach him

Area computer

Files

Switch matrix

Area computer

Files

Switch matrix

Fixed transmitter

Fixed transmitter

Fixed transmitter

Subscriber being called

Calling party

The caller 'dials' the subscriber's identification number and his home area code

Key:
Short-distance radio path
Land line
Switched public telephone network

Area computer

Switch matrix

Figure 20.2. **Switching between mobile radio transceivers.**

321

THE METROLINER
TELEPHONES

The high-speed trains (the Metroliners) between New York and Washington already have telephones. Some businessmen have stated that they travel on these trains rather than fly because of the availability of telephones. Calls can be placed either to or from persons on the trains. Any kind of telephone call can be placed, including credit card, collect (reverse charges), and data calls. When a person on the train is called, the caller must know the number of the train, which is listed in the train schedule. He does not have to know where the train is. The called party will be paged by a train attendant and asked to go to the telephone booth in his car.

The 225-mile track length is divided into nine zones of approximately equal length, and each zone has its own fixed radio transmitter (and receiver). As the train hurtles at 100 miles per hour from one zone to the next, calls in progress must be automatically switched from one fixed transmitter to the next. The switch happens without the customer being aware of it. As the train races into the Baltimore tunnels, the calls must again be switched to special transmitting and receiving equipment on the roof of the tunnel.

All calls go through the Bell System central office in Philadelphia where a special terminal remotely controls fixed transmitters. The terminal examines the noise-to-signal ratio from all the different transmitters and automatically transfers calls from one transmitter to another as the train rushes onward. When any available transmitter maintains a better noise-to-signal ratio than the one in use for several seconds, it is automatically switched in. The new transmitter is turned on 200 milliseconds before the old one is turned off. The user does not normally detect the change. The transmitter selected is not always the one closest to the train because of differences in the terrain or local conditions.

Frequencies in the 400-MHz (low UHF frequencies) range are used. A single frequency can be used more than once in areas geographically separate, as will be necessary with other mobile radio services. A telephone conversation may take place on one frequency near Washington while an entirely different conversation uses the same frequency near New York. The range of one frequency is about 100 miles, and two uses of the same frequency must be separated by at least 75 miles (see Fig. 20.3).

The automatic switching between antenna zones as the train travels and the reutilization of frequencies are demonstrations of two essential features of future mobile radio systems.

Nine zones along the track, each about 25 miles long

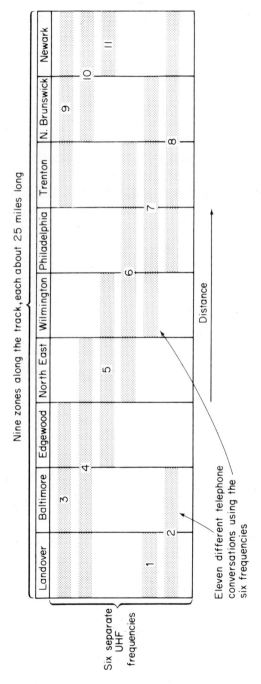

Six separate UHF frequencies

Eleven different telephone conversations using the six frequencies

Distance

Figure 20.3. Reutilization of frequencies for telephone channels on the Metroliner trains between Washington and New York. *Redrawn from reference 5.*

THE PORTABLE TRANSCEIVERS

In order for land mobile radio to achieve widespread usage, the users' portable sets must be reasonably inexpensive. Once the spectrum restrictions are lifted and appropriate fixed transmitters are in use, there will undoubtedly be a wide variety of different types of sets. At least one fashionable store in New York now sells a briefcase with a small radio-telephone in it (price $2550). A variety of portable radio transmitters and paging units are being used by construction workers, engineers, and others (Fig. 20.4). The computer-peripheral manufacturers sell portable radio terminals that are mainly used in vehicles today (Fig. 20.5). Figure 20.6 shows a science-fiction briefcase assembled by Honeywell; the only part of it that could not be mass-produced today is the television screen. The keyboards could be much more compact, leaving the business man space for his lunch! (Perhaps he will only have pills in the future!)

A mass market for inexpensive radio transceivers will perhaps provide sets in cars before it provides sets carried by people. If the market becomes large—anywhere near the rapidly rising curve in Fig. 20.1—then the cost of the mobile sets will be the largest portion of the overall system cost. Transmission and modulation techniques should be selected so as to minimize the cost of the set. For hand-carried sets, the low-power requirements necessitated by considerations of minimizing interference will be a factor in keeping the cost low.

COSTS OF TRANSCEIVERS IN VEHICLES

A report written for the President's Task Force on Communications estimates that the sets, using about 0.1 watt of power, would probably cost about $150 to $250 apiece in mass production lots [4]. Car radiotelephones, which pick up broadcast radio also and thus replace today's car radios, may add little to the car's cost if installed when the car is manufactured.

The New York City Transit Authority has a demonstration system covering 300 route miles of vehicle communications [4]. The cost of this system was $12 million. In a system such as that described above there would be additional costs for land lines, computers, and switching. The foregoing report [4] used the New York costs to estimate that a typical urban area with 2000 miles of highways and streets equipped with communications would require $42 million per year to cover the capital investment cost and maintenance. The calculation assumed that an average subscriber would make 500 calls per year and that the entire system would handle 375 million calls per year. The cost per call would then be 11¢. If the mobile transceiver costs $50 per year, this figure can be broken down to a cost of 10¢ per call, giving a total of 21¢ per call.

GRADE OF SERVICE
The preceding calculation went on to assume that the antenna sectors would be 2 miles long.

In the entire area there would therefore be 1000 antenna sectors covering the 2000 road-miles. The average city street was assumed to have four-lane traffic with 100 vehicles per mile per lane during heavy traffic periods. This figure gives 800 vehicles in one antenna sector.

It is assumed that, like telephone calls, the call duration from the vehicles is exponentially distributed. If the mean call duration is 3 minutes and the vehicles each originate one call per hour on the average, then the channel utilization

$$\rho = \frac{800 \times 3}{60\,M}$$

where M is the number of channels. This is 40 Erlangs of traffic.[1]

One hundred channels could be derived from one UHF television channel, assuming 60 MHz is used for each two-way radio channel and its spacing from adjacent channels. The result would be 0.4 Erlangs per channel.

More than one television channel may be needed to give overall coverage with adjacent antenna sectors, depending on how the radiation patterns are tailored. Not more than four groups of frequencies would be needed, as in Fig. 20.6. Today's umbrella coverage systems are estimated to cost about half the total cost of the transmitters in this system. However, they would provide only a small fraction of the numbers of channels if the same bandwidth were used. Only a small fraction of the vehicles on the streets could have transmitters, whereas here the aim is for all of them to have transmitters. Furthermore, the mobile transmitters on present systems need more power and are estimated to be at least twice as expensive [4].

In the preceding system, emergency services would not be subject to the queuing calculation. Firemen, policemen, and ambulances would have their own channels.

[1]For an explanation of "Erlangs" and the growing calculation, see the author's *Systems Analysis for Data Transmission*, Prentice-Hall, 1971.

USES OF VEHICLE TRANSCEIVERS

Telephone calls alone would make the car radio transmitter worthwhile for many people. Businessmen could set up meetings from their car on their way to work. Persons traveling could be in contact with their secretaries. Salesmen could talk to their clients while driving to see them; if caught in traffic jams, they could phone ahead rather than sit helpless. Highly active men presently waste much of their time in their car. With a telephone, they could employ some of that time.

The Touchtone keyboard would enable salesmen to transmit orders as soon as they were taken. On the way home they could use it to enquire how their stocks had fared. If lost, they could use their telephone to find the way. In fact, one channel on highways might be used for continuously giving directions.

The car need not stop at toll booths. A toll booth receiver could take its number and transmit messages for deducting the toll from the owner's bank account. Automatic toll collection may be used in some countries for regulating the flow of traffic into congested city centers. Parking fees may be similarly collected, and parking violations made expensive. One channel might be used for obtaining assistance in parking.

Car telephones could be used for a variety of safety functions as mentioned earlier, including fog and accident warning, breakdown assistance, and immediate help in a crash. It is possible that in the future radio transceivers will become a mandatory safety feature of cars, as seat belts and headrests are today.

SOCIAL IMPLICATIONS OF PERSONAL TRANSCEIVERS

The portable personal transceiver is probably further in the future than the car transceiver. However, once the spectrum engineering needs have been dealt with, portable sets may sell very rapidly, for they have many more applications than a car transceiver.

If they eventually come into widespread usage and the necessary computers are operational, it will be possible to locate a person almost anywhere in the country, provided that his transceiver is switched on and working. This situation need not necessarily constitute an invasion of privacy because he would always be free to switch off his transceiver or leave it at home. His home area computer might store the identifications of persons who tried to reach him when his transceiver was off the air.

As with other new telecommunication schemes, personal transceivers will probably be used by industry before a "consumer" market develops. Variations of this technology are now being used for paging individuals and for dispatching maintenance engineers. Key persons are likely to be made permanently accessible by means of a transceiver that they are not permitted to switch off during working hours. Calls to executives and other key personnel may be routed via fixed location secretaries, as today. It would often be useful for a secretary to be able to find her boss by radio.

Paging facilities can be connected to a private branch telephone exchange, as is the case with IBM's 2750 computerized PABX (installed and working well in Europe). With this system, a person called from a telephone extension can be paged under computer control. Public computer facilities that keep track of all subscribers will probably not be implemented quickly, although they could be built with today's technology. On the other hand, we may soon have computerized private branch exchanges that keep track of foremen, repairmen, or expediters in a factory or that provide "hot lines" between key individuals in an organization.

A portable radio in public use would be used for listening to broadcasts, as today, as well as for telephoning, obtaining time and date, and communicating with computers. The commuter on his tedious train ride could telephone home or arrange meetings in his office like the businessman with a car transceiver. He could fill in his time by dialing computers and using his Touchtone keyboard overlays as described in Chapter 11. He might learn French on the train from a computer, enquire after sports scores, or balance his checking and credit accounts. The radio facilities in the train would be a simple extension of what exists now in the New York–Washington trains. The call booths in the train would be replaced by a less expensive twin-wire antenna running the length of the train. In the earlier days of this technology, travelers without a pocket telephone might borrow one from the train attendant, with charges being automatically radioed for deduction from bank or credit accounts, as with everything else in this telecommunication-dominated society. The same facilities might be even more welcome on tedious airplane trips. Pan American might offer their passengers a radio chessboard for playing with computers!

The portable transceiver would give people a means of immediately contacting the police. This move might prove very popular in a society in which the growth rate of crime rivals the growth rate of electronics [6]. The police computers now spreading across the country would instantly pinpoint the emergency call and dispatch a patrol car. The fire and ambulance services could be contacted equally quickly. In a system with relatively small antenna segments, the approximate location of the transceiver sending an emergency call could immediately be found and a system could be devised so that precise positioning by triangulation is possible. Automatic positioning has been advocated for use in computer control of delivery trucks, taxis, "dial-a-bus" vehicles, and so on. Without automatic positioning, the emergency caller would have to give his whereabouts. Such a scheme would probably be a great deterrent to crime on the streets. One imagines the running criminal being tracked by the calls from persons near the scene of the crime as the patrol car closes in on him.

The portable transceivers may become an important part of the financial scene—a means of obtaining credit, for example. An individual might be securely identifiable by a combination of the transceiver he carries, which transmits a coded identification, and his "voice print," which is checked by a computer when he speaks his name over it. The microminiaturized circuitry could not be tampered with except by extremely expensive methods, and when used in conjunction with voice prints would be valueless to thieves unlike today's credit cards. Nevertheless, each person would guard his personal transceiver as he does his wallet and would report any loss immediately. Many applications of personal transceivers depend on whether the security problems can be solved, but the technology to do so now appears available.

The transceiver could also be used as a personal diary, shopping list, or memorandum. The user would have access to his own records stored in computer memory in voice form. His secretary or wife could store entries on his shopping list or diary.

Like all new technology, radio transceivers have potential for ill as well as for good. Bugging devices will be further improved. The police, for example, could arrange for automatic surveillance of persons possessing transceivers. Nevertheless, it seems likely that the social benefits will override the disadvantages.

The world of the ubiquitous transceiver may be two decades away, but sooner or later it will be as important to our society as the telephone is today.

REFERENCES

1. James Martin and Adrian Norman, *The Computerized Society* (Englewood Cliffs, N.J.: Prentice-Hall 1970), Chap. 10.

2. Report of the FCC Advisory Committee for Land Mobile Radio Service, 1968.
3. Eugene V. Rostow, "The Use and Management of the Electromagnetic Spectrum," the President's Task Force on Communications Policy, Staff Paper No. 7, Washington, D.C., 1969.
4. The President's Task Force on Communications Policy, Eugene V. Rostow. Staff Paper No. 1, Appendix F. "Concepts for Improving Land Mobile Radio Communications" by F. R. Eldridge, Washington, D.C., 1969.
5. C. E. Paul, "Telephones aboard the Metroliner," *Bell Laboratories Record*, March 1969.
6. James Martin and Adrian Norman, *The Computerized Society* (Englewood Cliffs, N.J.: Prentice-Hall 1970), Chap 5.

1909. Before the use of multiplexing or underground cables, masses of wires filled the streets. On Broadway (New York) some telephone poles were 90 feet high and carried as many as 50 cross arms. Courtesy A.T.&T.

21   COMPUTERIZED NETWORKS

There is much scope for computer assistance in using the new transmission networks. We have already discussed the added features in today's computerized telephone exchanges (Chapter 19) and the requirements for locating vehicles or persons with radiotelephones (Chapter 20). When we consider the additional uses of communications, other forms of assistance will be possible—in fact, in many cases, essential.

We are likely to see changes on telephone networks in private organizations before we see them on the public networks. The typical corporation tie-line system leaves much to be desired. It is constantly giving busy signals. Often it is difficult to track down a person on it, and it rarely handles data transmission efficiently. In the years ahead we shall probably see computer-controlled corporate networks, with their own switching and other facilities and, in an increasing number of cases, with international links.

The world's largest private telephone network is that used by the U.S. military—AUTOVON (*AUTO*matic *VO*ice *N*etwork). It has a total circuit mileage (in 1970) equivalent to the entire Bell System in the early 1950s. It handles voice and data communications, and calls can be encrypted if necessary. The circuits link U.S. military installations all over the world.

A corporate telephone network does not need to be nuclear-bomb-proof, but many other features of AUTOVON would be of value. A corporate network needs reliability, not "survivability." Like AUTOVON, it should handle data and voice over the same lines; also, as in the military, there will often be a separate specialized data network. As on AUTOVON, some calls will be highly important and should be connected as quickly as possible, but the majority of calls will be low-priority administrative traffic. It is desirable for certain top executives and their assistants to be able to contact each other immediately. On AUTOVON, all "command and control" calls are set up quickly, and this step could be profitably copied in industry. Today executives commonly receive busy signals from the tie-line network because the lines are flooded with unimportant chitchat.

PRECEDENCE CALLING

AUTOVON uses Touchtone telephones with 4 additional (red) keys to the right of the normal 12 keys (see Fig. 21.1). Pressing the P key designates the call as "priority," in which case it preempts ordinary calls.

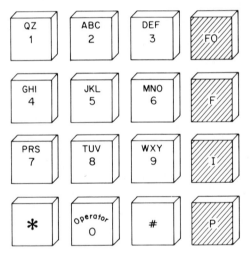

Figure 21.1. Four extra keys added to the Touchtone telephone keyboard to give precedence calls.

I (immediate) preempts priority calls. F (flash) and FO (flash override) give the highest levels of precedence. When a general presses FO, it overrides everything. The called party receives a special "precedence" ringing signal. A unique tone informs telephone users when they are being interrupted by a higher-precedence call. If one of these keys is used on a telephone not authorized to use that level of precedence, a prerecorded voice (what rank?) tells the caller that such a call cannot be put through.

HOT LINES In addition to the ability to make precedence calls, the general can have one or more "hot lines." He picks up a red telephone on his desk and is almost immediately connected to a predetermined location. Alternatively, he can press one key on a telephone with the same effect. The top executive in industry might have a "hot line" to his sales manager, distant plant managers, or presidents of subsidiary companies. In the future he is quite likely to have a hot line to an industrial "war room." The hot-line connections on AUTOVON are, in fact, switched. This factor makes them lower in cost and more reliable. A preprogrammed precedence level is automatically assigned to such calls, and service is so fast that a user often does not know that the call is switched. By the time he puts his ear to the handset, the phone of his hot-line mate appears to be ringing.

Another function of AUTOVON is the automatic setting up of conferences. A number representing a preselected list of conferees is keyed; the system looks up their numbers in its memory and switches lines to them. The conferees in this automatic hookup can be located all over the world.

In future private networks, a computerized PBX may be programmed to search for an individual, to try his number repeatedly if it is busy, or to page him by radio. Portable radiotelephones will undoubtedly be used by big organizations before they are used by the public. Initially they may function only within its premises; later they may operate within cities having appropriate radio facilities. The extension of the hot lines by radio will make key individuals permanently accessible.

In order to obtain these developments in private network switching, it seems essential that there be *competition* with the monopolistic common carriers. Left to their own devices, the carriers will rarely have any desire to change their switching practices. In private branch exchanges, such competition already exists where government regulations permit it. Where regulations do not permit it, there is probably little hope of change. We will discuss the issues of competition versus monopoly in the last chapter.

OBTAINING THE REQUIRED COMPUTER SERVICE Where terminals are used for communicating with computers, obtaining the required service may be simply a matter of dialing the right machine. As the variety of applications available from one terminal multiplies, however, computer assistance will be needed in finding the service. Furthermore, the terminal will often be on a leased line rather than on a dial-up line (at least with today's tariffs). The procedure may begin, then, by using the terminal to select the required service.

IBM has display units in its branch offices throughout North America connected on leased lines to its "Advanced Administrative System." After the user signs on, the computer helps him select the appropriate application programs with a "conversation" such as the following:

Terminal: FUNCTIONAL AREAS—ENTER LINE NUMBER:

1. FIELD COMMUNICATIONS SUPPORT
2. SALES AND QUOTA RECORD INQUIRY
3. DP ORDERS AND MOVEMENTS
4. OP ORDERS AND MOVEMENTS
5. SUPPLY ORDERS
6. CARD ORDERS
7. CUSTOMER ORDERS
8. PARTS AND PUBLICATIONS ORDERS
9. ACCOUNTS RECEIVABLE

User: 3

The user thus indicates that he is interested in the application area of "DP orders and movements." The computer then narrows down his field of interest further as follows:

Terminal: TYPE OF EQUIPMENT—ENTER LINE NUMBER:

1. SYSTEM
2. UNIT RECORD

User: 1

The computer then proceeds to establish what operation he wishes to carry out within the field of system DP orders and movements, as follows:

Terminal: TYPES OF TRANSACTION—ENTER LINE NUMBER.

1. PRE-AUDIT	8. CABLE ORDER
2. ORDER	9. CANCELLATION
3. RESCHEDULE	10. FUTURE RENTAL STOP
4. ALTERATION	11. INSTALLATION
5. SHIPMENT	12. RENTAL STOP
6. RECEIPT	13. INQUIRY
7. RPQ REQUEST	

User: 13

The computer has now determined which application programs to use to process the subsequent transactions.

In this system, all the application programs are presently in one computer. (Different computers serve different incoming line groups, but each has a complete set of programs.) In the future, terminals on leased lines in large corporations will access more than one computer and more than one data bank site. A conversation like the preceding one may be the means by which the switching computer connects the terminal to the processing computer it will use.

DIRECTORY COMPUTERS After another decade or so of development of computer usage, many different machines will be accessible by telecommunications for different functions. It is probable that some large general-purpose machines will offer many functions in the same system, but others will be separate, single-purpose systems, often handling a particular type of data base. Some computers will be highly specialized. There will probably be machines for setting print and assisting in editing, for example; these will be able to handle all the different type fonts found in modern books and will be able to set headings of different sizes. Book or magazine editors will work at terminals adjusting the text and formating the pages.

Gaining access to the different machines and their services may simply be a question of dialing the right number, which might be looked up in data processing "yellow pages." On the other hand, assistance in finding the right machine might itself be obtained from the terminal, and the telephone company might well have a directory-assistance computer for this purpose. A brief conversation using the terminal screen would identify the required computer, and switching to this computer would then take place. The directory computer might serve many of the common types of inquiries itself and then switch the user to another remote computer when unable to help. The remote computer, in turn, might itself decide at some point in the conversation that a different machine was required and thus arrange for switching to take place. Alternatively, it could merely obtain data or a program from the other machine and continue serving the terminal itself. Assistance from the telephone company switching computer would be of value in such networks; in particular, it would be of value in the billing of customers to identify which terminal loop originated the call.

In some cases, code conversion will be needed in communication between otherwise incompatible machines, and (as discussed in Chapter 10) such conversion may be a function of network interface computers. Clearly, many major developments in our transmission networks are needed in the years ahead.

INFORMATION
CONTROL
ROOMS

Obtaining information from its various computers will become a highly complex process in a large corporation, as many firms are now discovering in their efforts to set up "management information systems." The most successful formulas for such systems will be those that do not attempt 100% automation but use a judicious mixture of computers and human experience. One way to do so is to gather together the talent and terminals in rooms equipped to answer or redirect management's questions. Such rooms exist in embryo form today in a variety of organizations, and it is clear that they will grow in complexity and diversity of functions. They range from the "war rooms" of military command and control systems to centers for helping eliminate errors on data collection systems. In controlling a complex set of operations, the decisions that must be made are often brought together by telecommunications into one nerve center. A spectacular example is the NASA control center for the space missions at Houston, Texas. Many less-spectacular examples exist in industry.

The showpiece of many data processing installations in the future will probably be a "war room" in which all types of information can be sought from different remote computers and routed to locations where needed by management, salesmen, shop foremen, and others. One such center serving management at IBM's Armonk headquarters uses 16-mm film projectors, slide projectors, and closed-circuit television, as well as an array of different computer terminals. Generally, however, information systems better serve the needs of operational management and management making short-term decisions more completely than top management.

The functions of such information control rooms are likely to include the following:

1. The staff is responsible for the corporate data base from which management obtains much of its decision-making information. Errors and wrong information inevitably find their way into the data

base. There are many checks for detecting such errors, and when found, the control room staff is responsible for correcting them.

2. Some management questions are too complex to be answered easily from the terminals in their locations, and these questions are switched to the expert staff in the control room for answering. The control room staff can cause any display to appear on the screen of a manager's office and can monitor it on its own screen.

3. When emergency situations occur in the corporation, these situations may be referred to the control room, from which centralized expediting can be directed and the situation monitored.

4. When local management wishes to override a decision of its local computer, the management may contact the control room for arbitration. This procedure might be followed, for example, in a district sales office when the computer says that a certain order considered important by the sales office cannot be taken or met.

5. In some types of operations, computers should not be permitted to make *all* decisions in controlling them. The machine must be programmed to recognize when the intervention of a man with *experience* is necessary. The expertise for human intervention will be collected and made accessible in the one location.

6. The boardroom may also be equipped with terminals and screens and may perhaps be close to the "war room." This proximity would enable the highly specialized staff in the war room to be called in if needed to help in answering questions that arise during board meetings or to generate appropriate charts on the boardroom screens.

7. A hierarchy of information control rooms may exist. A factory may have a control room relating to the flow of work through the shop floor. A regional sales headquarters may have one relating to its customer and order situations. Subsidiary companies would probably have their own, separate from their parent company. A head office location may have a group of skilled operations research staff capable of dealing with more complex types of questions.

The manager of the future may have a hot line to his local information room. He may pick up a red telephone on his desk and be immediately switched to personnel who know where to find the answers to various types of questions. If possible, they will answer his question verbally, or display relevant information on the screen in his office, or prepare a printout for him. Sometimes they will route the query to another information room, perhaps more specialized, perhaps more expert, or perhaps in a far-distant part of the organization.

PUBLIC INFORMATION SERVICES

Since the earliest days of the computer, a trend from military usage of information systems to industrial usage can be detected. SAGE[1] brought real-time processing, visual display screens, and light pens; ten years later these devices were common in industrial systems. Now the technology of military command-and-control systems is being found in industry. Data transmission networks are spanning commercial organizations. The simulations used in "war games" are finding analogs in industrial and civil situations.

There may be a second stage to this trend. The schemes that were first used in the military at enormous cost reach industry when the cost falls substantially, and then in some cases reach the general public when it falls again. The data banks containing information of value to industry today will hold information sought by the public tomorrow. The public will use computers to help find a house, book theater tickets, provide stock information, give advice on optimizing tax returns, and so on. The terminals that spread from the military to industry will also spread from industry to the home. Just as the manager in industry needs help in using his terminal, so, to a greater extent, will the public. The lessons now being learned about the "man-machine interface" will be essential to home use of terminals; they are the key to acceptance.

Just as the information room in industry needs human skills as well as machines, so it would be a mistake in systems serving the public to attempt to automate everything. The public will need not only "directory assistance" but also many forms of more elaborate aid in using their terminals. As in industry, a human operator will often generate an appropriate display and switch it onto the screen of the inquirer.

In the early days of the telephone, the public received much more help from the staff at the telephone exchange than they do today. They could ask their "friendly operator" for the time, the weather forecast, how to get a nurse, or the name of the film at the neighborhood moving-picture house. Now they usually get a croaking recorded voice with a tone of monotonous indifference that is more likely to incense than soothe. Some countries still have friendly human help available by telephone. France, whose telephone system is notoriously underdeveloped, has a service called "Q.E.D.," which enables the public to ask questions about almost anything.

---

[1]SAGE means Semi-Automatic Ground Environment, the first of the systems designed to protect the United States from a surprise air attack.

In the early days of terminals for public use, it will be highly desirable to bring back the friendly human operator. When the average person is using the terminal to shop, to book a journey or a vacation, to balance his bank or credit account, to search for literature on a particular subject, or for calculation facilities or any number of other functions, he will occasionally need assistance. The computer rooms providing such services will need a staff for this purpose. The terminals should have a HELP button. Information rooms with skilled personnel similar to those in management assistance will be needed in many other areas of terminal usage. One prime usage will be in obtaining the switched connections to the requisite computers. Let us end this chapter, then (in a book mainly concerned with automation), with a plea for the skilled use of *people* in the systems. The trend of employment toward service industries and away from the manufacturing industries will continue in the United States, and one segment that will need people will be that of the rapidly growing information services.

One of the most difficult sections of the American coast-to-coast telegraph line was built over the Sierra Nevada Mountains in 1861. On the early telephone lines, 40 years later, it became possible, amazingly, to speak across a distance of 2000 miles, from New York to Denver, without amplifiers (because the vacuum tube had not been invented). Courtesy of The Western Union Telegraph Company.

22 THE INCREASING BANDWIDTH

As mentioned earlier, the transmission bandwidth being using both for radio and for cable has been steadily increasing. Work in the development stage at present indicates that these bandwidths will go on increasing for the next decade. It was thought that we would reach an upper limit with radio beams on earth because of absorption, interruption of the beams by aircraft or birds, and because of the physical dangers from higher-frequency beams. Now, however, it seems clear that infrared and optical transmission are going to be used, giving us "through the air" beams of much higher frequency than conventional radio. By using cables, waveguides, and laser pipes, we will be able to progress *much* farther up the electromagnetic spectrum than at present.

Figure 22.1 plots the capacity of major telecommunication links since 1840.

In 1819 Oersted discovered the relation between magnetism and electricity, and Ampere, Faraday, and others continued this work in 1820. Shortly afterward, the first systems were prepared for transmitting data in which information was conveyed by the slow swinging of magnetic needles (page 68). A 40-mile telegraph line was set up between Baltimore and Washington by Morse in 1844. The information was coded in dots and dashes, and a steel pen made indentions in a paper strip. In 1849 the first slow telegraph printer link was set up, and by the early 1860s speeds of about 15 bits per second were being achieved. In 1874 Baudot invented a "multiplexor" system, which enabled up to six signals from telegraph machines to be transmitted together over the same line.

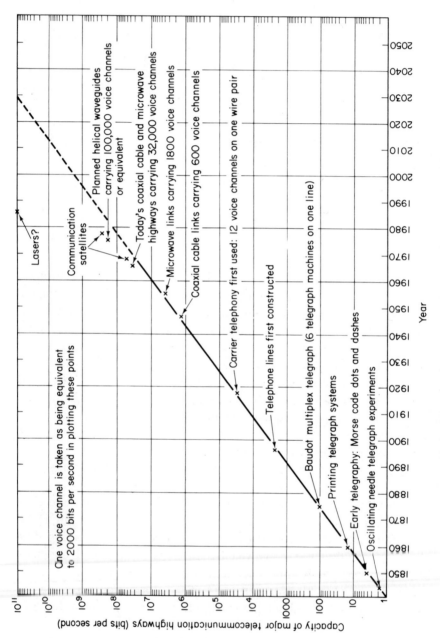

Figure 22.1. The sequence of inventions in telecommunications.

Telephone lines were first constructed in the 1890s. In 1918 the first "carrier" technique was used, and 12 voice channels were sent over one wire pair. In the 1940s coaxial cables were laid down to carry large numbers of voice channels, and in the 1950s microwave links were built. Today, in the United States, a single long-distance telephone cable or microwave chain commonly carries 11,000 voice channels, and already cables with more than twice this capacity are in use.

The remarkable fact about Fig. 22.1 is that the plot gives a straight line for such an extended period of time. One is immediately led to speculate that the straight line will continue into the future; perhaps it will continue a long way into the future. After all, it has been straight for 130 years and we still have oceans of electromagnetic spectrum unused as yet. It is possible that by the year 2100 we will be modulating gamma rays.

Certainly we have evidence that channel capacity is going to increase for the next ten or fifteen years. Helical waveguide channels are about to be laid. Satellites using multiple millimeter-wave beams are being designed. We know that the laser can be modulated to give very high bit rates, and we suspect that we are far from unlocking its full potential as yet. Indeed, the activity in today's development laboratories suggests that, if anything, the curve will swing upward above the straight line.

It is important to note that the vertical axis is drawn logarithmically. If this diagram were drawn with a linear axis representing capacity, and the entire past history occupied the height of one page of this book, then the next seventy years would need a scale about one mile in length.

The straight line in Fig. 22.1 is going up by a factor of 10 every 17 years. Oddly enough, the number of radio transmitters in use is also increasing exponentially and has been going up by a factor of 10 every 17 years. In 1950 there were 3 transmitters per 1000 persons in the United States. In 1967 there were 30 transmitter per 1000 persons, somewhat less than half of which were mobile two-way radios. The President's Task Force on Communications Policy forecast a continuation of this exponential trend, in which case there would be 300 transmitters per 1000 persons by 1984. Most of these transmitters would be mobile radios.

In this chapter we will assume that the straight line in the figure is going to continue for another thirty years, and we shall attempt to outline possible consequences. An engineer will protest that he cannot see any mechanism that will make this growth possible. An accountant will argue that he cannot see any economic justification for the prodigal expenditure.

Others said the same thing in 1880, 1920, 1950, and at numerous other times, but the straight line continued.

We will also assume that computer logic circuitry continues to become cheaper, smaller, more reliable, less power consuming, and less heat dissipating—in other words, that the promise of LSI circuitry is fulfilled. We will assume that the size of on-line file systems continues to increase and that their cost continues to decrease. It might be incorrect to extrapolate the curves of Chapter 16 too far into the future, and so we will assume that they continue for another fifteen years. Laboratory evidence seems to support this prediction. After that, we will assume, the growth continues but at a substantially slower rate.

Given these assumptions, we will outline a table of events that should be regarded not as a forecast but as a statement of what is likely to be possible. Whether or not discrete events in the table come into being will depend on the entrepreneurs of the day, on the laws, and on the societal structure. The reader might choose to slide some of the events listed backwards or forwards in time. One could state a number of devices or uses of technology today that are technically possible, economically reasonable, and socially desirable but have not come into being. However, basic technological trends like increasing bandwidth will almost certainly continue, partly because of a desire to explore farther in space but also because of a desperate struggle for military superiority and preparedness. When the great technical nations are threatened with nuclear or biochemical extinction, technology will progress.

It is certainly desirable to forecast the potential of technology in a world that is rushing headlong into changes so powerful that the whole fabric of society is likely to be altered.

NOT SO MUCH A FORECAST AS A STATEMENT OF POTENTIAL

Early 1970s

Cable TV spreads rapidly. Cables capable of carrying 40 to 60 channels are installed in many areas. Cable TV companies increase their own production of local TV programs. Cable TV is extensively used for educational television and quality commercial-free movie reruns.

The TV cable is used experimentally for services other than television, including hi-fi music distribution, computer-assisted instruction in the home, and use of the television set as an interactive terminal capable of digital use or of displaying color images.

In North America, Picturephone is marketed extensively to business. Picturephone switching is available initially only in a few major cities but is rapidly spreading to other cities. A few affluent homes have Picturephone sets.

There are isolated uses of the Picturephone links for very high speed data transmission.

The marketing of machines for playing movies in cartridges over television sets begins and grows rapidly. The Columbia Broadcasting System's EVR (electronic video recording) on film is the first, but is soon challenged by alternative (incompatible) systems.

Touchtone telephones are installed extensively throughout North America. Only sets with 12 keys are installed.

Telephone voice answerback by computers spreads rapidly. A wide variety of terminals employing voice answerback are marketed. There is general recognition of its usefulness in creating an effective man-machine interface, especially in applications where data are being entered by the terminal user as in factory data collection and sales-order entry systems.

Use of the Touchtone telephone as a computer terminal spreads, for diverse applications. A variety of devices are marketed to connect to the telephone for this system, including additional keyboards, cheap strip printers, a terminal identification device for security purposes, and an attachment for displaying responses on the screen of a domestic TV set. "Voice-print" recognition is used on an experimental basis on some systems for identifying a terminal user. Credit cards with a coded, machine-readable, magnetic strips come into use.

Switched data networks are installed, initially serving only certain cities but then spreading rapidly. Different types of systems are used in different countries, which makes the international marketing of terminals and communications software more difficult.

Broadband data transmission over public, private switched, and leased point-to-point lines becomes widely used.

Computerized telephone exchanges, such as ESS 1 and 2, are installed at a rapid rate. Computerized private branch exchanges are also installed in large companies.

Facsimile mail and document transmission increases in popularity and drops in cost. Conventional mail delivery in cities such as New York increases in unreliability. It becomes clear that facsimile gives the only way to deliver mail rapidly and hence a large market grows.

The Bell System installs many miles of helical waveguide transmission systems, each capable of carrying more than 230 thousand telephone calls simultaneously. As yet there are limited locations that have sufficient traffic volume to justify installing this.

PCM links are widely installed and terminals for transmitting data over them come into use. Time-division multiplexing of the PCM links provides many data channels. The maturing of "large-scale integration" circuit technology gives rise to many equipment designs using digital transmission.

Much larger satellites are launched into synchronous orbit. These satellites have much greater power and capacity than Intelsat III (the generation of satellites at the time of writing). Less expensive and more numerous earth stations are used. Most underdeveloped nations now have worldwide links via satellite.

Cable TV systems begin to use satellites for distribution of their programs.

Educational television is transmitted directly to village antennas in India from a NASA satellite, and it becomes clear that this step can have a major impact on the education of the underdeveloped world.

Satellites for domestic use are launched on a pilot basis in the United States. They are used both for direct television broadcasing to community antennas and for common-carrier system trunking of all types of signals.

Late 1970s

Picturephone service, at a substantially reduced rate, is now available on a public dial-up basis in many urban locations. An increasing number of affluent homes are installing Picturephone sets. In cities offering the service most *major* offices have at least one Picturephone set. Many have a picturephone room with facilities that any authorized employee can use. Some have a picturephone conference room with sets capable of switching transmission between the faces of different conferees. In addition, many executives have a Picturephone set on their desk. The dial-up Picturephone lines are used for very high-speed data transmission.

By 1980 3% of all business telephone sets are Picturephone and 1% of all domestic sets (AT & T forecast). In certain high-income urban and suburban areas, the majority of homes have Picturephone service, which has become a necessary status symbol. Telephone callers receiving no picture say, "Goodness, aren't you on Picturephone yet?"

Video-conferencing facilities using television links rather than Picturephone grow in usage. Television gives higher-quality pictures and can display documents so that they are readable. Most countries outside North America have not started a Picturephone service but give much publicity to their few video-conference links. Video-conferencing rooms, giving very attractive facilities with voice-actuated camera switching, are installed in many major cities. Large corporations use their own video-conferencing facilities with leased television links. In some cases, executives can participate in the conferences without leaving their office.

Cable TV continues its rapid growth until more than 65% of American homes have this facility. In spite of the name, many segments of these TV links are by radio rather than cable. In the cities, low-power microwave radio is used to feed the cable, often taking the signal to the tops of apartment blocks. In the country, some local transmitters are fed by the cable, and these take the signal to remote homesteads.

Much television becomes "local" in its programming rather than national or regional. On the other hand, much of its becomes international because of the increasing capacity for satellite distribution and the desire to fill up more of the many channels on the cable. England's BBC sends many of its intelligently produced studio programs abroad. When the British television channels close down about midnight, the evening's programs are transmitted to America, where it is then prime time; and when America goes to bed, *its* programs are transmitted to Japan and Australia.

High-fidelity music is piped into the home on a request basis in cities having enough subscribers to make a "music library" service profitable. Cable TV channels are used for this purpose. In some areas, live performances, as well as the music library tapes, are piped into the home.

Wall screens, 5 foot and larger, without cathode-ray tubes, are much demonstrated. A high-fidelity television service using the big screens is planned. Many authorities advocate that it be confined to cable TV rather than to broadcasting because it would consume a very high bandwidth. There are now many new demands other than television for the VHF and UHF frequencies.

Cartridge movies become a mass market. The machines for playing them are used extensively in industry and schools as well as in the home. Cartridges with still frames are used, along with many microfilm devices, for providing manuals and training courses. Movie libraries and rental services are used by the consumer market but not as extensively as was hoped. It is thought that "piped" movies into the home may have a deeper market penetration.

Domestic satellites become a basic part of the North American common-carrier networks. A second type of satellite can distribute television to antennas small enough to be installed on rooftops. These are generally used as community antennas, but some are installed in individual homes, especially in isolated areas.

The number of satellites broadcasting to the lesser-developed nations increases. The cost of appropriate television sets and antennas has dropped substantially. The sets are used for education, entertainment, and propaganda. Although television in the United States continues to focus on entertainment programs, because advertising revenue supported it, television in the poorer nations often does not, for it is largely under governmental control. There is little or no advertising revenue, and thus the entertainment programs are broadcast mainly to lure viewers into watching the education or propaganda broadcasts. The satellites used are American and Russian, and many thousands of villages in India, the Middle East, Africa, Asia, and South America are seeing television for the first time.

The Touchtone telephone keyboard with 12 keys is now almost universally available in North America. Eventually the Bell System's production of rotary-dial telephone ceases.

Data processing continues its prodigious growth. Computers have dropped in cost greatly, and with the "fifth generation" they promise to invade almost every aspect of life. In spite of the proliferation of mini-computers, the fastest-growing area of data processing is still that of communication-based systems. Conversely, the fastest-growing area of telecommunications is still data transmission. With the spread of terminals displaying pictorial images and with real-time intercommunication between computers, data processing promises to consume enormous quantities of bandwidth if it can be made available cheaply enough.

Dial-up data networks offering a wide range of bandwidths spread in many countries, and are profitable. Their incompatibility problems have to be solved with interface computers. The tariffs in the United States make the cost of data transmission independent of distance, other than for very short distances. This factor has had a major effect on the organization of data processing in nationwide corporations, which, in turn, has an effect on management structure. A corporation commonly maintains one nationwide data bank for almost all kinds of functions—rather than regional or local files.

After a long legal battle, "foreign attachments" are permitted on Picturephone lines through a rigorously designed Bell System interface. A wide proliferation of devices making use of the high dial-up bandwidth come on the market. These devices include color picture sets, video-recording equipment, and high-fidelity sets using signal processing to encode only *changes* in the image and using 10 or 30 frame changes per second with techniques to eliminate flicker. There are devices for the remote reproduction of engineering drawings, often from a master stored on microfilm or in digital form in computer files. A variety of facsimile equipment, remote Xerographic, and photoreproduction equipment exists. A host of computer peripherals use the Picturephone lines.

Digital transmission is now accepted as the way of life for the telephone companies for both long- and short-haul trunks. New telephone, Picturephone, and television trunks are all digital. Bit rates recommended by CCITT become standard and typical are those of the T1, T2, and higher carriers.

There is now extensive use of computer-assisted instruction. A minority of the programs used are exceedingly effective. These programs gain prominence, and the vast mass of ineffective programs are largely forgotten. The style and technique for developing good teaching programs become understood.

Many high-IQ families in North America have some form of computer terminal in the home, often employing the keyboard of their telephone and the screen of their television set. Teaching children becomes one of the major motivations for home terminals.

An increasing amount of white-collar work is done in the home through the use of terminals and, sometimes, Picturephone sets, particularly in the case of businessmen, programmers, persons writing computer-teaching material, and those who develope the vast amount of material needed in the numerous data banks. Some software firms chiefly employ married women working from the home. Typing services and text editing services also operate from the home via a typewriterlike terminal connected to a central computer and its files.

Crime increases appallingly. Many homes and offices have a burglar detection apparatus connected to the telephone lines so that the police computers are notified automatically of any intruder. Radar techniques are being experimented with for computer identification of intruders. Pedestrians begin to carry radio alarms for calling for police help.

Privacy and security of information in computer data banks become a major concern but not before the occurrence of many serious scandals and crimes. Substantial development expenditure makes it possible to maintain very high security of data transmitted and stored. New laws are passed in some states relating to the use and control of personal data stored in computer systems.[1]

The first serious attempts at radio spectrum engineering have made possible a greatly increased use of car radiotelephones. Many vehicles are now equipped with a Touchtone keyboard. However, many more measures are needed in spectrum engineering to meet the ever-growing demand. Political pressures are clearly preventing an optimal use of the radio spectrum.

Early 1980s

The cost of long-distance calls has dropped still further, and now the cost of international calls has fallen dramatically. Many parts of the world can be telephoned from New York at only slightly higher cost than telephoning Chicago. The satellite links are used to handle data, facsimile, and television with equal facility. This fact, together with supersonic travel and the worldwide spread of American hotel chains, makes the planet seem much smaller.

The drop in cost of long-distance transmission in the United States had a major effect on the organization of national corporations and their information processing. Now the same effect is being felt on international corporations. Data banks of

[1] See James Martin and Adrian Norman, *The Computerized Society* (Englewood Cliffs, N.J.: Prentice-Hall, 1970), Chap. 24.

international corporate data are used by large firms in the United States, Europe, and Japan. As in the military establishment a decade before, decisions in the field could be instantly flashed back to centralized command posts. Factories are sited where labor costs are low. Laboratories and programming centers are set up where talent is plentiful and cheap. Administrative offices are located in countries with favorable tax laws. All are linked with leased lines, and now data dial-up costs are becoming favorable because of international data networks.

Data banks with 10^{14} bits of directly accessible storage are fairly common. Such storage is used for photographs, drawings, and documents in image forms, as well as for digital data. Some data banks store randomly accessible film sequences. Much telecommunication usage is for access to the numerous data banks.

The "war room" in business is now conventional. It takes many forms and is given many different names. Often it is the showpiece of a firm's data processing. Few offices of top management are complete without their video link to the firm's information center. There is now (after some bitter failures) a general recognition that the human element in the information center is as important as the machine element. Experienced and highly professional staffs operate with an array of terminals and wall screens that often rivals a NASA Mission Control Center in appearance. Although some managers like to demonstrate their prowess at operating their own terminal, many have an assistant for this task, or else they use their video link to a local information room, which in turn may route some questions on to a remote or central information room.

A major change is occurring in the way financial payments are made in the United States. Cash transfers take place *within* the electronic systems. "Electronic fund transfers" take place within one computer, or between two different computers, holding the accounts of the persons concerned. They are sometimes initiated by "preauthorization"—that is, instructions in advance of the payment being given to the bank's computer for the routine payment of salary, rents, dues, etc. They are sometimes initiated by the use of an electronic fund transfer card—a development of the credit card that contains machine-readable details about the holder and which is inserted into an on-line terminal in stores, restaurants, and offices. The cost per transaction is substantially lower than with checks or credit cards. There is no longer any talk of a "checkless" society. The number of checks in use has risen to 50 billion per year and bankers desperately hope that electronic fund transfer will lessen the deluge of check processing. The EFT terminals have done much to prevent the high level of crime that became associated with credit cards.

Telecommunications is being extensively used in medicine. Information from all manner of patient instrumentation is transmitted to specialists or computers. "Prediagnosis" interviews are carried out between patient and distant computer, often to determine whether the patient should see a doctor or not, or visit a hospital.

Automatic monitoring of chronically sick patients is done by computer, sometimes with the automatic administering of drugs. Sometimes patients are monitored during normal daily activities by means of miniature instrumentation connected to radio transmitters (as with the astronauts). In some cases their readings are recorded by a tiny machine that they can later link to the telephone and transmit the readings to the hospital computer. Remote diagnostic studios are used with powerful television lenses. With the help of a nurse in the studio, a distant doctor or specialist can examine a patient as though the patient were in his office. The patient can see him and talk to him. The doctor can fill the whole of his color screen with the pupil of the patient's eye, or tongue or skin rash. He can listen to a distant stethoscope and can see both instrument readings and computer analyses of them.

The Picturephone tariffs drop further in cost; the service spreads rapidly, with business becoming increasingly dependent on it. It offers many peripheral services in the home and is now regarded by some as a necessity rather than a luxury or status symbol. Air and other forms of transportation became steadily more congested and unpleasant throughout the 1970s, and new efforts to reverse this trend are barely succeeding.

Computer voice input systems that permit a user to speak to a computer over the telephone, using a very limited vocabulary of clearly separated words, are marketed. The computer responds with spoken voice words.

The first public laser channel for common-carrier operation comes into service, giving a digital channel capable of carrying several hundred thousand telephone calls. A small flexible laser cable using highly refined fiber optics, comes into use. It's bandwidth is lower than the common carrier laser transmission systems but it is an ideal replacement for the TV coaxial cable in the home. It provides a cable into the home that could carry 500 wall-screen channels.

A "high-fidelity" television service is started with a large number of lines to the screen. Five-foot wall screens are marketed and the sets operate in a digital fashion. The digital bit stream using a form of differential modulation reaches the home over a coaxial cable with frequent digital repeaters. It is not planned to transmit high-fidelity television over VHF or UHF radio.

Schemes employing millimeter-wave radio frequencies for signal distribution come into public use in some cities. The repeaters are spaced at frequent intervals down the streets or on the city rooftops. A very high channel capacity with a low level of noise and distortion is made possible. The highly directed beams eliminate interference between different transmitters, and it is clear that a large number of such systems using the same frequency bands will be employed in a city. Infrared and optical transmission links are used extensively between rooftops, especially for the transmission of data. Satellite systems using millimeter wave frequencies are employed with great success. High frequencies appear essential because of the congestion in the microwave band.

Late 1980s

People, at least in North America, are becoming accustomed to a society in which many functions of life are carried out by telecommunications. Numerous persons now work at home at least part of the time. Tax deductions for home facilities needed for work are standard and include computer terminals, Picturephones, and video-conference screens. Many firms install these devices at their own expense in the homes of employed persons who need them. Many homes are now built with a childproof, wifeproof office. America's traditional antipathy to soundproofing diminishes.

There are now many millions of fairly inexpensive electronic fund transfer terminals in use. Some persons have one in their home.

A new generation (of people) is now dominant which can communicate with the computers with ease over the various transmission links. Programming is taught at an early age in schools, and most well-educated persons under 30 can use one programming language fluently. The computer and software industries have spent much time and money developing the "man-machine interface" so that the ubiquitous terminals are usable by the greatest number of people. Nevertheless, some minds seem naturally at home with the new technology, whereas for others it is a struggle. Some persons seem to have a built-in hostility to this form of communication, which is becoming so vital in society.

A person who is well adapted to the technology can carry out an amazing number of different functions from his home terminals. An ever-increasing world of computers, data banks, sound, film, and picture libraries is there for him to explore. Many authorities, however, still believe that the technology is only in its infancy. Certainly a vast amount of work lies ahead in building up the data banks, writing computer teaching programs, improving computer-assisted medical diagnosis, and so on. Many data-bank uses that met with initial skepticism from the professional men they were designed for, are now gaining wide acceptance, but the work required to make them comprehensive is enormous.

Much television interviewing takes place remotely with the subject being in his office or home, and interviewer being in the studio many miles away, often in a different country.

"High-fidelity" television proves to be popular, and cables for it are laid down in large numbers. Some affluent homes have wall-sized screens. The vividness of the large color picture provides a more "hot" medium in McLuhan's sense of the word than the earlier small TV screen.

It is clear that there is a grave shortage of channel bandwidth (as at most other times in history!). Higher-capacity cables are laid into the house. Millimeter-wave systems are used for carrying digital TV down the streets. The capacity of the nationwide helical waveguide channels is increased, and many thousands of miles of the new laser channels are built. Domestic satellites using laser frequencies are launched on a pilot basis, with the intention of building a domestic network that employs a large number of such satellites.

Three-dimensional television is demonstrated, using large wall screens.

"Personalized" newspapers come into operation. Instead of being presented with an impersonal and often superficial selection of all the news, as today, a subscriber may register his news-requirement "profile." He then receives detailed news on topics that interest him. This information may be printed out on his home terminal or stored for him so that he can display it on his screen when he wishes. All types of categories can be registered—for example, local news about a district other than where he lives, news about a particular industry, company, or stock, scientific news, movie reviews by particular reviewers, news about crime, sex, war, or business, and foreign news.

Powerful satellites now permit home pickup of worldwide television with rooftop antennas. Some television programs are dubbed in many languages. Because a television sound channel requires only about one thousandth of the bandwidth of the picture, this does not substantially increase the overall bandwidth requirements. A rooftop antenna can now pick up almost as many television programs as sound stations on the shortwave radio of the 1960s.

The use of computers as a hobby has by now become widespread, and a major section of industry has grown up to cater to the computer amateurs. Although some have mini-computers, the majority gain access to the machines with large files, using the various telecommunication facilities. Magazines for the amateur market have come on the scene, and many systems and data banks have been designed for their use. They can obtain newsletters at their terminals and can register a "profile" to determine what categories of information they receive. Computing spreads like a drug to a large number of people, and once hooked they cannot let the machines alone. Computer amateurs, much more so than the radio amateurs of an earlier era, are able to make significant and original contributions to the industry, especially in programs and in data bank contents.

Telephones in vehicles are widespread. Cars are manufactured with the option of a Touchtone keyboard on the dial, alongside all the other electronic equipment. Discussion exists on whether a car telephone, plus a loudspeaker that is never switched off while the vehicle is on the public roads, should be a legal requirement. The loudspeaker would receive tones and voices that interrupt the car radio or music player and that are concerned with safety, accidents, parking, toll paying, route-finding, and so on. Improvements in the nationwide dialing system for vehicles are needed. In many parts of the country, a car still cannot be reached by a long-distance telephone call.

Personal portable telephones have been in use for some time by the military forces, who have achieved almost worldwide dialing of key persons. In the cities, fire, police, and other personnel have portable radiotelephones. Many corporations use portable telephones within factories or office buildings, and some are now setting up a nationwide corporate dialing network. A public service that is an extension of the car telephone service and that permits small transceivers to be carried anywhere is now clearly practicable. A nationwide paging system has been in use for some time, but there is no facility for the paged person to respond directly. He has to go to a telephone. There is pressure to make paging illegal in some areas in order that an individual there can feel free from the relentless pursuit of the new communication facilities.

Early 1990s

"Direct-access" data banks now exceed 10^{18} bits where usable. "Read-only" bulk storage is much lower in cost than storage that can be updated.

Mass production and mass marketing have given most homes wall screens. Laser channels are being laid into homes to give a very large number of high-fidelity TV channels. Dial-up channels (as opposed to cable TV channels) are being installed in limited numbers to carry the large-screen pictures. The latter are being sold mainly to industry and, to a lesser extent, to affluent homes. The situation regarding large-screen, dial-up facilities is about the same as that with Picture-phone's in the mid-1970s. The office of a top-ranking executive is now likely to have a wall-sized screen facing his desk. It can either be connected in its entirety to another location or fragmented into several smaller screens for conference or multimedia use. All such facilities are digital. Office and apartment blocks have small computers that act as concentrators and control screen fragmenting and switching.

The wall screens are now frequently used by lecturers. A lecturer may conduct a class of 16 or 32 people, all in different locations and all using Picturephone sets. They may be in their homes; they may be in worldwide business locations. The lecturer can see all of their faces, and they can see either his face or diagrams, objects, slides, or film clippings, which he switches on to their screens. He can speak to any of the "class" individually and they can speak to him. They cannot see their fellow students and so tend to ask questions with little embarrassment from possible class reaction. The teacher may occasionally let his students see the rest of the class. If he wants, he may switch the face of a questioner onto the class screens. On the other hand, he can address any one student without the others hearing. Similar facilities are used for sales meetings, management briefings, and by a manager addressing his employees who work at home. Some such links in industry are international.

Printed newspapers stop production in the United States except for a minor intellectual press, a few picture newspapers for low-IQ readers, and some local newspapers.

Television news has become exceeding vivid. At least one cable TV channel carries big-screen news all day. Very high quality photography on the wall screen shows riots and world catastrophes in fine detail, including local wars (limited nuclear or biochemical) and a new wave of famines in certain underdeveloped and overpopulated countries.

On the other hand, the unrest in the underdeveloped world is undoubtedly inflamed by their television window to the affluent nations.

A directory the size of the "yellow pages" is published, listing all the telecommunication services available and giving the codes necessary for communicating with the innumerable computers. Most shopping is done from the home screens. With increasing automation, employment in the service industries has become much greater than in the manufacturing industries, and now many persons are employed in giving specialized assistance in the information networks. The HELP key on the Touchtone telephone overlays (Fig. 11.10) and other devices are used frequently. As in the information centers of industry a decade earlier, it is found that much more useful facilities can be provided if specialized human assistance and human expertise are built into the system. A most useful attribute on many computer systems is an on-line human being. Almost 10% of the United States Gross National Product is spent on the communication services discussed in this chapter (including broadcasting).

Late 1990s

Most people now carry a portable radio transceiver with a Touchtone keyboard. They have a wallet full of credit-card size overlays. When an individual is dialed, he can be reached in most parts of the country. The zones of radio inaccessibility are diminishing. It has been suggested that the public should be issued with transceivers that transmit their national identification number, even when switched off. These devices would help in controlling crime, which is still growing at an appalling rate. They would also be used in most financial transactions.

Picturephone is now being rapidly replaced with a dial-up service that uses the digital wall screens. The wall screen can be linked either to the TV cable or to the new dial-up video-phone network. With the latter, one apartment or one room in a house can be connected to another via the wall screens. Large-screen videotape recorders are inexpensive, and families often dial up relatives to play them videotapes of the children.

The dial-up channel makes it possible to request movies for individual playing in the home.

"Personalized" video news services are set up. In the same way that printed news became individually selected for each subscriber "profile," now video big-screen newcasts are similarly assembled.

In addition to being able to select what he wants instead of passively watching what is fed to the masses, the home viewer now has a channel on which he can "converse" with the medium. This channel is often used for teaching. "Computer-assisted instruction" thus progressed from terminals giving an alphanumeric response from the computer, to terminals giving a color slide response and presentation, and then to terminals giving short movie sequences. All these facilities in turn were initially used in institutions and later were a dial-up or cable TV service to the home. The interactive media are used for many functions other than teaching, but much "passive" programming is still used also.

In large cities, the movie libraries are located close to the telephone (video-phone) central offices; and as the channel is permanently connected from the central office to the home, the telecommunication cost of dialing for movies or news is not high. In rural districts, however, it is very high because of the long-distance transmission to the movie library. Here the distribution of movies in cartridges is likely to continue. In spite of the vast growth of telecommunications, there are thus still advantages in metropolitan living. The high bandwidth telecommunication services are much cheaper in the cities. We have not yet reached the point when the drop in long-distance transmission costs can overtake man's ability to consume increasing bandwidth.

Public movie theaters have declined under the competition from the home entertainment media and now cater to two main markets. First, they show movies with a degree of sex and obscenity not permitted on the home media and catering to a low-IQ market too poor to afford the wall screens. Second, other theaters show spectacular movies on screens occupying 180° to 360° of the field of vision; these screens can create an impact greater than the home wall screens.

Circular-domed rooms come into use in which the entire ceiling and walls are a three-dimensional color TV screen. These can be linked by a communication line to any appropriate camera system, videotape, or transmitter, or they can be operated from their own cartridge, which can generate an environment of sylvan tranquility, spring in the Andes, earth orbit, or the Kiluaea volcano erupting.

The ability to program and communicate with computers is now widespread.

Among persons under 30, the inability to program is regarded as a form of illiteracy. There is a serious gap between the capabilities of young and old people in communicating with the all-pervading machines.

There is experimental use of drugs administered under electronic control in conjunction with entertainment media, largely to heighten and "edit" emotional reaction.

It seems clear to many authorities that the staggering advances in molecular biology are going to merge with the electronic technology. This prospect is dismaying to many older people (those who read this book in the 1970s, but strangely enough, a new generation of students is emerging that appears to welcome it!

SECTION V

LAW AND POLITICS

The International Telegraph Union and related bodies have done much to achieve international standardization. This photograph shows the heads of national delegations attending the first conference in Paris, 1865.

23 THE LAW AND POLITICS

Technological change has reached a fast and furious pace and is rapidly becoming faster. Most of society's institutions, on the other hand, can only change slowly, and in many cases avert their eyes from the frightening onrush of science. Herein lies a dilemma that will make itself felt in chaotic ways in the decades ahead.

Some uses of technology in society will involve increasingly large expenditures of money and exceedingly elaborate systems design. This prospect is true for the new schemes for urban transportation, for example. It is true for air traffic control. It will be increasingly true for medicine, educational facilities, civil engineering, urban renewal, city center design, pollution control, environmental design, ecological problems, and many computers uses. It is especially true for telecommunications.

Where vast expenditures and public issues are involved, the arguments can rapidly become ideological rather than pragmatic. Some countries will insist on government ownership and civil service control. In others, ownership design and management by the private sector are an inviolate creed.

The trouble with telecommunications, as with some other public services, is that segments of the industry need to be monopolistic. Having more than one telephone company operate in the same city with competing lines and switching offices is too wasteful and is likely to result in inconvenience for the users. In telegraphy, for example, some American users have had to pay a high price for having Telex machines on Western

361

Union lines side by side with TWX machines on AT & T lines, solely so that they could reach different subscribers. Furthermore, where limited physical resources are involved, as with uses of the radio spectrum and satellites, it is necessary to employ these resources as efficiently as possible. These factors make telecommunications a "natural" monopoly, at least in certain aspects, and they have led to state ownership in some countries and tight government regulation in others. Unfortunately, both state ownership and regulation have led to inefficiencies and to the sapping of incentive, sometimes to a degree that has been highly destructive. And so we have a problem.

Today, with the diverse new technologies discussed in this book, the ideal approach would be that of a massive piece of systems engineering. The telephone lines, the Picturephone service, satellites, data links, mobile radiotelephones, and cable TV could all be interlinked in one gigantic system that would use the different facilities to their best advantage. Tight radio-spectrum engineering would ensue, and there would be an optimum balance between use of radio and use of cable for different functions. The old systems and plant would be integrated into the new. Computers would control the multipurpose switching and the tracking of mobile system elements. The result would be massive engineering design on a scale almost comparable with that of the moon shot.

It is a version of this dream that lies behind the state ownership and control of telecommunications in many countries. Another aspect of the subject is that the telecommunication facilities are vital for national defense and security.

The 1969 bill in England creating the British Post Office Corporation gives it a monopoly over all potential uses of telecommunications, including the provision of channels for broadcasting, data transmission, mobile radio, remote reading of electricity meters, burglar alarms, and so on. The corporation has the "exclusive privilege" of "running systems for the conveyance, through the agency of electric, magnetic, electromagnetic, electro-chemical and electro-mechanical energy . . . speech, sounds, pictures, signals to control apparatus and signals serving for the importation of any matter otherwise than in the form of sound or visual images." This privilege includes the regulation of communication between "persons and persons, things and things and persons and things." As the London *Economist* said when the corporation was formed, that seems to include everything up to love at first sight.

The corporation even notified the companies renting television sets for cable connection that it could, conceivably, take over their rental in 1976 when there will be a general review of broadcasting.

The British government's approach to the future of telecommunications is summed up in a statement to the House of Commons by John Stonehouse, the Postmaster General. In view of our comments in Chapter 2 about technological surprises and forecasting failures, his statement seems a classic of doctrinaire socialism: "We must ensure that all possible technical developments are caught within the monopoly . . . *before they are invented* and before they become competitive with the monopoly, for that would be a very embarrassing position." One wonders how many inventors will save Mr. Stonehouse embarrassment by bringing their invention to the United States.

The United States also has its dogma. An essential part of the American creed is that public communications networks should be run by private industry. The opening section of the President's Task Force on Communications Policy (Final Report) says, "We have taken pains to protect society against the risks of concentrated power, in the hands of either government or of the communication companies." And later, "In approaching these problems, we have been guided by the basic premise underlying the law and policy affecting American industry and commerce: that, unless clearly inimical to the public interest, free market competition affords the most reliable incentives for innovation, cost reduction, and efficient resource allocation. Hence competition should be the rule and monopoly the exception."

An extensive section on "Future Opportunities for Television" laments, as indeed it should, the lack of diversity in American television. The average American family has its television set switched on for more than 6 hours per day. Most sets can obtain many times more channels than sets in most other countries. Yet a foreigner visiting the United States is frequently amazed that such an affluent country has such poor-quality radio and television, strangely lacking both diversity and quality. The opportunities offered by the medium have been gravely missed. The Task Force report discusses several possible solutions to this problem, including pay-TV, cable TV, and various sources of educational TV. However, it makes *no mention whatsoever* of the only solution that has given diverse and intelligent programs in some other countries—a nationwide government-operated network like the BBC in Britian or the NHK in Japan. This total omission seems to suggest that the topic is taboo in the United States.

If private enterprise has failed to provide quality radio and television programming, it has certainly succeeded in providing the world's best transmission network. One might judge the relative claims of government versus private ownership of telecommunications, or of competition versus absolute monopoly, by grading the performance of the two systems. Most government-controlled monopolies would not score very well. Indeed, most would fail abysmally on many counts. Countries with such systems generally have long waiting lists for telephones, woefully inadequate data transmission facilities, out-of-date switching offices, expensive long-distance calls, hardly any portable radio transceivers in public use, and regard the Picturephone as science fiction. Broadband data transmission is virtually unknown. One hears horror stories of people waiting 2 years for a telephone to be installed.

At the end of the 1940s America began to replace step-by-step (Strowger) switch telephone exchanges with the faster, more flexible crossbar switch. In the late 1960s it began to replace the crossbar-switch exchanges with vastly superior computerized switching (Chapter 19). Britain is still installing Strowger-switch exchanges (although dabbling with a very small and apparently unprofitable electronic exchange).

As noted in Chapter 4, the British Post Office has forecast the start of a switched video-phone service in 1990—20 years after America. France and many other countries with state-run telecommunications are worse off than Britain.

It seems that competitive forces are essential to keep the telecommunication companies on their toes. Monopoly becomes moribund. The problem, then, is how to preserve the integrity of the vast network, avoid excess duplication, and conserve scarce natural resources, such as spectrum space, without stifling invention and entrepreneurship. Let us examine a few of today's major issues.

1. International circuits

There are two main ways to provide a circuit today across the ocean, satellite and submarine cable. The cost per circuit is far lower with satellite than with cable; yet suboceanic cables are still being designed and laid. Why?

The satellite delay time of a quarter of a second has been cited as a reason by the cable proponents. The added military security that comes from having more than one transmission medium is another reason. However, the most important fact is that today no single firm in the United States is permitted to operate *both* cable and satellite or whichever it chooses. We therefore have vast vested interests lobbying for cable on the one side and satellite on the other. The present investments are likely to be regulated on the basis of compromises that seem fair to industry claimants. Without a change in the institutional structure, they seem unlikely to be made on the basis of minimum social cost.

One INTELSAT IV satellite costing a few million dollars could handle all the present Atlantic basin traffic, and a second one could handle all the Pacific traffic. Yet the FCC recently authorized a new trans-Atlantic cable, TAT 5, which will require an initial investment of $70 million and will require revenues of $250 million over its useful life [1]. A cable to the Virgin Islands was also authorized to operate in parallel with the satellite. The cables have only a small fraction of the satellite capacity; yet the FCC stipulated that new traffic on the Virgin Islands cable was to be allocated on a ratio of 50:50 between satellite and cable. The cost to users of such decisions will be high.

The great hope that satellites bring is that the cost of international traffic will drop to a small fraction of its present cost. With the large satellites ahead, this prospect is likely. However, if satellites are forced to compete with cable on a 50–50 basis, the dream will not materialize. At the time of writing, more than two-thirds of Comsat's original capital remains in cash or securities rather than in productive communication assets [2].

The President's Task Force, reviewing this situation, argued strongly for the formation of a single entity for all United States international transmission. This entity would operate cable and satellite circuits, or any other that seemed appropriate.

The Task Force also considered the possibility of unregulated competition in this area but concluded that such competition was unreasonable because it might eventually result in the elimination of cables; it also argued that it would be disadvantageous to have more than one organization launching the satellites. Furthermore, establishing conditions for effective competition would be very difficult. The report concluded that "a natural monopoly is rarely encountered in the real world" but that the transmission segment of international communications appeared to be one.

The creation of the "single entity" should rationalize this industry. The Task Force went on to recommend, however, that the entity be subject to conditions limiting the monopolistic disadvantages, as follows [3]:

1. It should be limited to the function of providing the transmission links and should sell only transmission capacity. It is possible also that the earth stations could be user-owned. The transmission capacity would be sold to competing common carriers. The entity would thus be a "carrier's carrier."
2. It should not engage in manufacturing or have manufacturing affiliations. These factors should be provided by the competitive marketplace. Satellite and earth station manufacturing would remain a highly competitive industry, thus ensuring vigorous and rapid technological development.
3. It would not provide domestic U.S. transmission or have affiliations with domestic carriers. This prohibition would ensure diverse development of domestic satellite potentials.
4. "It would be subject to strengthened government regulation."

In spite of the Task Force's well-developed arguments, it is far from likely that the "single entity" will be formed by Congress. Both AT & T and IT & T will fight to avoid losing their cables.

There can hardly be an international satellite monopoly owned by a single nation, and therefore it is important that any institutional change strengthens the concept of INTELSAT. President Johnson in his message to Congress in 1967 setting up the Task Force [4] reaffirmed strongly America's support of INTELSAT. He stressed that America sought no domination of satellite communications and urged the Soviet Union and nations of Eastern Europe (but not China) to join INTELSAT. The message concluded: "The challenge of this new technology is simple—it is to encourage men to talk to each other rather than fight one another. Historians may write that the human race survived or faltered because of how well it mastered the technology of this age."

2. Domestic satellites

The economic desirability of using domestic satellites is rapidly increasing. If Comsat had been permitted to launch them, we would almost certainly have had domestic satellites by now. In 1970 President Nixon recommended that the FCC permit "Any financially qualified public or private entity, including government corporations" to launch and operate communication satellites. These could be for their own exclusive use or for common-carrier use. He suggested that this policy be adopted for an interim 3- to 5-year basis. The President made it clear that he regarded competition as the best ultimate course, which seems a radical departure from the philosophy of Comsat or AT & T.

There are many technical groups who would oppose this recommendation. The President's Task Force, after detailed study of the problem, stated that a pilot scheme was necessary to develop expertise and test such questions as what radio interference would result. It suggested that there should be broad participation in the pilot program, although one member disagreed, saying that Comsat should be solely responsible. He based his view on the "unfortunate" experience with divided ownership of international earth stations.

Competition will probably give rise to a greater diversity of schemes and therefore may give the best results in the long run. Its danger is that the industry might become locked into a less-than-optimal system, and political pressures might prevent it backing off to a more integrated approach when the technology is better understood. The integrated approach could eventually bring major economies of scale.

The Ford Foundation suggested the NASA would be the best organization to develop pilot satellite schemes. With the present cutback in the space budget, such development might be a desirable way of using some of NASA's talent and facilities. If they were as successful as with the moon shot, we should have a good satellite system. However, it seems unlikely that NASA *will* control a pilot project.

The debate could continue endlessly: Which is best, free competition or integrated systems engineering á la NASA?

3. The development of cable TV

As indicated in Chapter 5, cable TV offers a wide variety of exciting prospects. The existing TV broadcasters, however, have opposed it so vehemently that its development was seriously retarded. Considerable resources can be commanded for such fights. In 1966 the FCC was persuaded to impose rules so stringent that they virtually stopped extension of cable service. In the top 100 markets (about 89% of all TV sets), they required operators who wanted to import distant signals to prove that doing so would not harm any existing stations or *any that might later be established*.

More recently, the FCC seems to have repented, at least partially, and now most cable operators are permitted to bring in the three major networks, some independent channels, and one educational channel. In addition, they have been authorized to originate their own programming and to sell advertising for it. The FCC also ruled that they can use short-haul microwave links for distributing their signals. The show, at last, is on the road.

While the cable casters were under crippling attack from the established

368 THE LAW AND POLITICS

networks on the one side, the telephone companies attacked them on the other. The short-haul microwave ruling is one victory scored against the telephone companies. Cable TV fits naturally into the telephone companies' service pattern, and it has the great appeal for them of being a business without rate regulation. The cable TV companies needed to have space on the telephone company poles, and often the telephone company manufactured the cable equipment. In leasing the equipment to the cable TV companies, telephone company contracts frequently prevented them from originating programming and from engaging in two-way communications. The prevention of two-way communications rules out most of the interesting applications of cable discussed in Chapter 5. The FCC is now attempting to protect the cable casters from the massive telephone companies.

The preceding case is an example of how the regulations constraining free enterprise retarded rather than encouraged the development of exciting new facilities. The problem seems to be that the regulations and attitudes of the regulators did not evolve as rapidly as the technology.

4. The UHF television channels

As we discussed in Chapter 13, there is now a desperate radio spectrum crisis. Many valuable new uses of radio are being prevented because of spectrum starvation, particularly in land mobile radio applications.

The frequencies usable for such applications are from 5 to 890 MHz. In the United States, the federal government reserves about 300 MHz of this amount for itself in total. Of the remaining 585 MHz, no less than 420 MHz are allocated to UHF television broadcasting—the active band from 470 to 890 MHz. This figure is in addition to the 72 MHz of VHF television frequencies (channels 2 to 13) in normal use.

This allocation was made in 1949. Unfortunately, UHF television in the United States (unlike Britain) has never really got off the ground. Few commercial stations broadcast UHF at all. One reason is that the transmitting equipment is more expensive than VHF. Until 1964 very few sets were able to pick up UHF, and then a law was passed saying that all new sets must be able to pick it up. Still, many sets today cannot pick it up, and those that can are rarely used for this purpose because there are no clearly marked UHF channels that the channel-selection knob clicks to.

The rapid growth of cable TV at present makes it seem even more unlikely that UHF will take off; and in view of "the silent crisis" of the radio spectrum, the correct action when all factors are considered is surely to distribute new television services by cable rather than use radio frequencies, at least in the denser areas of the country where cable TV is economic. At least some, and possibly all, of the unused UHF frequencies should be freed for other purposes.

The powerful television industry, however, is fiercely opposed to this view and is capable of great political pressure. The UHF allocation costs broadcasters nothing, and even if they do not use it they refuse to give any of it up to other types of users. The FCC, perhaps feeling guilty that it enticed many investors into UHF, where they lost money, generally supports the television industry. It considered extending the ban on distant signal importation for cable TV on the theory that UHF growth requires shelter from distant signal competition [5].

Here again we see powerful vested interests, backed by the regulatory authorities, impeding a very important new technology.

5. Foreign attachments

In 1968, after a bitter legal fight, the FCC ruled that AT & T must permit "foreign attachments" to its system. Most of the world's telecommunication organizations prohibit the attachment to their public network of "foreign" devices from other manufacturers. They claim, as did AT & T, that it is necessary to maintain the "integrity of the network" and that signals from foreign attachments may interfere with the switching or multiplexing processes.

An enormous variety of devices *could* be attached to the telephone network. The possibilities seem endless in computer terminals alone, but the devices could only be attached via a relatively expensive telephone company "data set" of limited characteristics. It is important that inventors be free to design and sell any device they think of, provided that it does not, indeed, interfere with the correct working of the network. We need inexpensive devices that can be used in the home. We need portable terminals that can be used in a public call box, and there are innumerable possibilities here with LSI circuitry. We need new types of telephone-connected burglar alarms, perhaps using radio links. The scope for invention is enormous. However, in most countries, no one invents anything to attach directly to the telephone because of prohibitive laws. The telephone companies themselves are slow to introduce new consumer devices. If the "foreign attachment" laws did not exist, there would soon be devices for attaching hi-fi units, electric cookers, light pens, electric power switches, and goodness knows what else to the public networks.

The FCC, still laboring under the obsolete Communications Act of 1934, ruled that the prohibition of foreign attachments was "unreasonable, unlawful, and unreasonably discriminatory." Now, at last, devices can be attached through a protective connecting arrangement. An inventor still has to use the telephone company's dialing and hook switch equipment, but this limitation is also under attack. As a result of the new ruling, the electronics magazines were soon full of advertisements for new devices, including modems. An interior decorator can, if he wishes, now design his own telephone.

This type of change in the law seems vital for progress. In many countries, archaic and monopolistic laws prevent anything other than telephone company machines being attached to the networks. Often it is illegal to use even machines with no direct electrical contact (acoustical or inductive coupling). Many computer personnel think that a technical reason is responsible for this, but there is no technical factor that cannot be surpassed. The reason is a legal one and it is doing great harm.

A country with aspirations for technical progress must make sure that anachronistic laws and hidebound regulating bodies do not stand in the way.

6. Interconnections with private networks

Somewhat similar to the question of foreign attachments is that of interconnections with private communication systems. Many organizations have their own communication system for data or voice. It is often advantageous for such systems to be interconnectable with the public network. It is also desirable for radio operators to be able to "patch" their radios to the telephone lines. It is desirable that leased-line networks to computers, often employing concentrators or multiplexers, be publicly dialable. Such interconnections have been illegal in the past. Now, in the United States, the interconnection rule has been eased. As for the future, it would be worthwhile to have all the different systems discussed in this book interconnectable.

An historic ruling was made by the FCC in June 1968 that determined much of the future of foreign attachments and interconnections. In 1965 Thomas F. Carter filed an antitrust suit against AT & T because the latter threatened to prevent him from marketing his "Carterphone device," which enabled mobile radio systems to be connected to the telephone network. The eventual ruling opened the door to such attachments.

7. New networks

After the Carterphone decision, a second precedent-shattering development took place in the United States in August 1969 when the FCC authorized a small company, Microwave Communications, Inc., to set up a public microwave system between St. Louis and Chicago, paralleling AT & T's routes. John D. Goeken, the 37-year-old president and founder of MCI, had spent 6 years and $400,000 fighting a legal battle with the telephone company to obtain this right. He said he could offer a wider choice of bandwidths than AT & T and prices as much as 94% cheaper.

After the FCC decision, Goeken suddenly found massive sources of finance available to him and applied to extend the system nationwide. Within a few months the FCC was deluged with 750 other applications for new networks. The plans of the University Computing Corporation to set up the Datran System (Chapter 10) were boosted. The large common carriers are spending much money to fight the MCI ruling, however.

Clearly there must be a limit to the amount of such competition that can be permitted. The difficult question is where to draw the line. AT & T could have marketed facilities like MCI's or Datran's; however, they did not. They were protected from the need for such innovation by the law. The new competition is a much-needed shot in the arm.

A major part of the problem is that new companies want to "skim the cream" of the telecommunications market. The large common carriers are priced to furnish their services to remote as well as to metropolitan regions. It is expensive to run telephone lines to houses in sparsely settled regions, and the common carriers make up their losses in such areas by appropriately charging in the dense regions where the cost per subscriber is low. If a new company competes only in the low-cost regions, then it can naturally charge much lower prices than the big carriers. The eventual outcome of unregulated competition would be that the telephone charges in remote areas become very high.

Some would contend that this situation should be permitted to occur and that an artificial price structure is wrong; a remote farmstead should pay heavily for its telephone. Are we going to do the same with cable TV when it provides services that over-the-air TV does not? Should the TV cable to the remote farmstead be subsidized? Should Picturephone and data transmission be cheaper in the 50 or so largest cities?

8. New switching facilities

The new rulings governing foreign attachments and interconnections with noncarrier networks are a major spur to innovation and entrepreneurship. One could imagine that telephone and other networks as channels with standard interfaces so that inventors could design devices which conform to the interface, like plugging a lamp into a wall socket. A mechanism for controlling such inventions is necessary because there is a serious danger of their interfering with other subscribers or even damaging the network. The telephone company, however, could even be saved installation costs if standard plugs were provided. When a subscriber moved his apartment he could take his telephone with him and plug them in. We are still a long way from this situation, but it seems an ideal to legislate toward.

The question of innovations in switching is quite different. There are many new facilities that one would like in public switching offices, some of which were discussed in Chapters 19 and 21. These facilities cannot be introduced by a private entrepreneur, only by the telephone company. How can we make sure that we have a sufficient level of inventiveness here? The problem is difficult. If the telephone company wants to leave its switching offices unchanged for 40 years, it can do so. It is immune from competition in this area.

Perhaps the best solution would be for the FCC to press for changes when they are thought feasible. Is it too much to hope that instead of regulating against innovation, a government agency should ask, "This is a good idea: why are you not doing it?"

Left to itself, the regulated monopoly is going to find many reasons for avoiding change. The telephone companies are prepared to keep their latest design of switching office for the next 40 years, although in a world of intercommunicating computers and real-time terminals the design may be hopelessly inadequate. The countries with government-provided telecommunications have almost all fallen so far behind in the provision of data channels that they have seriously impeded the otherwise rapid growth of the data processing industry.

In the rapid change of the next three decades (Chapter 22), it is going to be highly worthwhile to be constantly spurring the telephone monopolies; otherwise the rich potentials of our new-found technology will not mature fully.

9. Line broking

Except in special cases, the sharing of leased lines by different customers has been prohibited in the United States and most other countries. It has rarely been possible to set up as a "line broker," leasing a broadband channel and employing one's own equipment to offer services on that channel to different customers. Recently in the United States there has been some relaxation of the sharing and line broking rules.

The ruling against sharing lines did not do much harm in the days when we were only concerned with telephones. Today the line-sharing prohibition retards the development of several important functions. An entrepreneur should be able to take the MCI offering of microwave bandwidth and set up services for its customers utilizing it. A firm offering data services would be free to set up multiplexing facilities, concentrators, and store-and-forward switching for its customers, because with data these facilities can give *much* more efficient line utilization. A firm should be able to set up a "hot potato" network (Chapter 10) with lines leased from the common carriers, if it wishes. Many large corporations have leased-line networks with their own private switching facilities, and these switching facilities often have functions that differ from the public switching offices. It would make sense for more than one firm to share such private facilities. It should be permissible to go into business to provide such networks, using leased lines. Doing so would be one way to introduce competition into the provision of switching facilities.

10. The reward system

The U.S. common carriers have their profits regulated by the state and federal governments in a system known as the "reward system." A complex formula is used, and to some extent it determines the attitudes of the carriers. The choice of formula can either impede or encourage technical progress.

Several aspects of today's reward system seem to be impeding technical progress. The carriers are rewarded on the basis of return on investment. The operating expenses, including depreciation of plant, are subtracted from the operating income. The resultant net income determines the maximum return, which means that carriers are encouraged to build up their plant. The greater the book value of the plant, the greater the allowed return. Carriers are encouraged to make the operating expenses as low as possible, including maintenance and depreciation. Therefore they build long-lasting plant and depreciate it as slowly as possible. It is normal to design telecommunications plant for an operating life of 40 years. Long-lived plant with little maintenance cost earns the maximum "reward."

A 40-year replacement cycle might have appeared reasonable in the 1930s, but in today's electronics world the rate of change is so fast that equipment is obsolete almost before it is installed. Few persons in data processing expect to keep a computer longer than 6 years. An IBM 7070 computer which cost nearly a million dollars in 1964 was sold by the Parke-Bernet Antique Galleries for $2250 in 1970. The purchase price of computing equipment is commonly equivalent to about 3 or 4 years' rental; yet most commercial machines are rented. The Bell System ESS computers for switching, however, were planned to have a 40-year operating life.

The promise of today's technology and the demands of the data processing world would imply a rapid installation of PCM, computerized exchanges, Touchtone dialing, and so on. The reward system, however, encourages carriers to replace existing equipment only slowly.

Fortunately, current expansion of the networks is rapid, and therefore new types of plant *are* being installed fast. If the carriers double their circuit mileage every 5 years, for example, then in 20 years equipment in use today will account for only about one-sixteenth of the total network. In this case, the problem is one of interfacing the old equipment with the new. The compatibility problem is a constant drag on new development. The electronics of 1980 telecommunications circuits may have to be compatible with some installed in1940 if the rules of the game do not change.

A more rapid replacement of equipment would need larger capital investment; hence faster progress has a price tag on it. Perhaps the new services rather than the old telephone service ought to pay the added cost. Because of the rate of expansion, the cost is not as high as it would otherwise be. If only one-sixteenth of the total network is over 20 years old, it would seem unreasonable to scrap everything over this age that does not fit into the latest methodology.

The general breakdown in the New York telephone service at the time this book is being written seems to be due in part to obsolete and ill-maintained plant and to shortage of attention to the new user class who transmit data. Some would say that this situation is a direct result of an inappropriate reward system. It is also said that the reward system is responsible for the common carriers unprogressive stands on satellite questions, interconnection and shared-use rules, as well as other issues.

11. Radio spectrum allocation

One of the most pressing problems is the allocation of the radio spectrum. It seems clear that flexibility is needed in spectrum management. The present method of allocating frequency bands nationwide is out of date. Spectrum engineering and allocation by TAS (time-area-spectrum) packages (Chapter 13) are needed. Transferable "property rights" in spectrum may be a solution. The President's Task Force recommends an "eclectic" approach introducing major—if incremental—modifications of existing administrative, economic, and engineering practices.

It summarizes its conclusions as follows [6]:

A. As a basic guideline, we should seek that combination of spectrum uses which offer maximum social and economic contribution to the national welfare and security.

   Accordingly, the following principles emerge:

   1. We should seek the continuing substitution of higher-valued spectrum uses for lower-valued uses and the *addition* of uses whose net effect is to increase overall benefits, with due consideration of all imbedded capital investments.

   2. Unused spectrum resources should be employed to meet any legitimate need provided that this does not cause excessive interference to existing uses, conforms with established standards and international agreements, and does not interfere wih established plans for higher-valued uses.

   3. Comprehensive coordination of all spectrum use is required, under a continuing framework of public administration.

B. Greater consideration of economic factors is necessary.

   1. An improved schedule of fees for spectrum licenses should be developed, which reflects the extent of spectrum use (e.g., bandwidth, power, service area, time availability and the level of demand for spectrum rights. And intensive studies should be conducted of other means to account for economic value, including adjustable license fees, spectrum leasing, and taxation.

   2. License privileges should clearly be stated for each class of spectrum use (e.g., land mobile, radio relay, etc.) in terms of interference probability, channel loading, service quality, and other appropriate factors.

   3. Administrative procedures should be modified to permit greater trans-

ferability of licenses among legitimate spectrum users within broad service classifications, subject to all relevant conditions of the initial license, including the requirement that all exchanges or transfers be registered and approved by the spectrum management authority.

4. Procedures should be developed whereby a prospective spectrum user may obtain a license even though this would represent a potential source of harmful interference to an established clear channel user, provided that prior arrangements are concluded between all affected parties, including adequate compensation or indemnification by the new user.

C. Greater attention to individual spectrum uses should be achieved through "spectrum engineering" and related technical considerations.

1. A more flexible approach to spectrum management should be adopted, under which the National Table of Frequency Allocations is transformed over time from a fixed allocation by user category to a basic planning guide by service classification.

2. A comprehensive spectrum engineering capability for individualized planning and engineering of spectrum uses should be developed, charged with continuing improvement in technical design and operating standards for all transmitting and receiving equipment and other devices that materially affect spectrum use.

D. Enhanced management capabilities and a restructuring of responsibility and authority are required.

1. Legislation should be considered which would vest in an Executive Branch agency overall responsibility for ensuring efficient spectrum use for all government and non-government uses; this legislation should contain appropriate guidance as to coordination required between the spectrum manager and the FCC in areas of mutual interest and concern.

2. The agency should be given the resources needed to develop a strong interdisciplinary capability embracing technical, economic, social, and legal skills, to support its spectrum planning, management, and coordination responsibilities as described in this Report.

3. In particular, the agency should: (a) determine and continually update the division of spectrum among various classes of users, and administer its use on the basis of detailed planning and engineering, at local and national levels; (b) establish and enforce technical standards applicable to all transmitting and receiving equipment and other devices that materially affect spectrum use; (c) coordinate federal R&D activities oriented toward spectrum management and use, except those directed to fulfill a specific mission of another agency; and (d) administer any user fee system now existing or later established.

4. In the interim, to meet existing spectrum management problems and to prepare for the future, resources should be provided to begin effectively to implement the general and specific recommendations of this report.

E. Specific recommendations in selected problem areas.

1. Land mobile radio services

   (a) Land mobile radio services should be authorized to use spectrum resources now within the national allocations for UHF television broadcasting which are unusable by television stations under the present TV station allotment plans; subject to operating criteria which will avoid harmful interference to television broadcasting on adjacent channels or in adjacent geographic areas.

   (b) Equipment and operating standards should be established for engineering future land mobile services to permit closer spacing of base stations sharing the same frequency assignment: the use of multi-channel radio equipment should be encouraged wherever this would economically provide more efficient spectrum use.

   (c) Development and use of common-user and common-carrier mobile radio systems—including those employing wire-line trunking between individual base stations—should be encouraged, particularly for users with intermittent service requirements.

   (d) A range of channel loading criteria should be established to encourage effective frequency sharing among complementary uses and to provide a satisfactory and well defined quality of service to each user.

   (e) The sub-allocation of land mobile spectrum bands by user class should be substantially discontinued. Any remaining sub-allocations should be flexibly administered within each geographic area.

   (f) Procedures should be established whereby members of the general public now restricted to the Citizens Radio classification may be licensed to use certain land mobile spectrum resources subject to compliance with reasonable technical and operating standards and appropriate channel-loading criteria.

2. Public safety radio services

   (a) The public safety radio services, in particular, should be incorporated into the government spectrum allocation and management framework.

   (b) Operating standards requiring greater time and geographic frequency sharing among public safety agencies should be established.

   (c) Development of common-user mobile radio systems for public safety services should be encouraged.

3. Television broadcasting

   (a) Spectrum resources presently allocated for broadcasting which are unusable for that purpose under existing station allotment plans should be made available for land mobile and other uses.

   (b) Studies of improved techniques for television broadcasting should be carried out on a continuing basis, with respect to alternative distribution methods, channel bandwidth reductions, and reduction in total spectrum allocations.

4. Microwave service (1000–10,000 MHz)

   (a) Improved operating standards (e.g., modulation, antenna directivity, space diversity, etc.) should be established to achieve greater spectrum re-use and interference protection between terrestrial facilities sharing the same frequency ranges.

   (b) The criteria for satellite/terrestrial sharing of all spectrum allocations below 10,000 MHz should be re-evaluated, giving due consideration to the significant technical differences between domestic and international satellite systems and to improvement in technical data since the existing criteria were established.

   (c) Experimental programs should be conducted to ascertain the probability of interference between satellite earth stations and terrestrial radio relay stations, in shared frequency bands below 10,000 MHz.

   (d) Improved criteria and coordination procedures should be developed for efficient sharing of spectrum alloc tions and orbital locations among domestic and international satellite systems, both government and non-government.

5. Millimeter-wave bands (above 10,000 MHz)

   (a) Continuing research and development activities needed to bring about effective and efficient use of these spectrum bands should be encouraged, through federal R&D programs and flexible policies with regard to the potential uses of these bands.

   (b) Existing domestic allocations of all millimeter-wave bands should be reviewed to determine the feasibility of inter-service sharing of these bands as an alternative to exclusive domestic allocations to satellite and terrestrial services.

GOVERNMENT FUNDING

Looking at the eventual future of a telecommunications plant, it seems certain that it will become fully digital, with digital switching. Analog signals will be carried in digital form, and *large-scale integration* logic will be used throughout the network. The cost of transmission with such a network will become a fraction of today's cost, and the capability of the network will be much greater.

A problem exists in how to progress from today's analog plant to tomorrow's digital plant. To make the transition it may be necessary to make investments that will make the operation of today's telephone plant temporarily more expensive, as shown in Fig. 23.1. Taking course B rather than course A will result in lower costs in years ahead, at the expense of higher costs in the immediate future. However, if course B represents the building of a digital network with telephone, data, and other facilities integrated, it will provide much greater capability than course A in the years ahead. The value of this to the economy can be

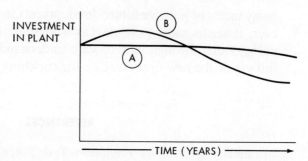

Figure 23.1. Which nations
will take path B.

estimated only very approximately but is likely to be immense, even if one discounts intangible benefits such as computer-assisted instruction in the home (which might become technology's most powerful societal force).

Future value to the economy, especially when filled with so many intangibles, rarely motivates businessmen to open their purses. If Wall Street investment bankers select the course, it will not be course B. Failure of imagination is absolute if there is not a high return on venture capital in a few years.

To develop atomic energy, massive government funding was necessary. This has been true in other high-cost technologies such as aircraft and space. In a sense it has been true with automobiles if one counts the cost of highways, and a car without roads is like a Picturephone set without communication lines. The payoff for using government funds to push digital telecommunications over the hump in course B will be greater than the payoff from peaceful uses of atomic energy, the moonshot, or the Anglo-French Concorde.

Some nations probably will spend the money to take course B. Many governments, however, will lack the imagination and the courage to invest sufficiently in this vital future resource.

**THE BATTLES
AHEAD**

It seems clear that great battles are impending in the field of telecommunications. The stakes are high and the changes needed are revolutionary. The fighting will be fierce.

When the battles are over, the vast interlocking networks will look very different from today. The condition of man, living in a maze of electronic signals, will become profoundly changed.

There is a danger that politics, lobbying, monopolistic sloth, or destructive competition will rob us of part of the riches that the technology could bring. It behooves our politicians and regulators to understand fully the many facets of possible future developments in telecommunications. However, it seems sadly possible (as with some other public uses of technology) that through lack of such understanding, we shall fail to make full use of the new opportunities for enriching man's world.

REFERENCES

1. Final Report of the President's Task Force on Communications Policy, Chapter 2, page 9, Washington, D.C., December 1968.
2. Communication Satellite Corporation annual report, 1969.
3. *Ibid.*, Chapter 2, pages 38–47.
4. Message to Congress from the President of the United States transmitting recommendations relative to world communications. House of Representatives Document No. 157, 90th Congress, 1st Session, August 14, 1967.
5. "Future Opportunities for Television," Chapter 7 of the Final Report of the President's Task Force on Communications Policy, Washington, D.C., December 1968.
6. "The Use and Management of the Electromagnetic Spectrum," Chapter 8 of the Final Report of the President's Task Force on Communications Policy, Washington, D.C., December 1968.

GLOSSARY

There is little point in redefining the wheel, and where useful the definitions in this glossary have been taken from other recognized glossaries.

A suffix "2" after a definition below indicates that it is the CCITT definition, published in *List of Definitions of Essential Telecommunication Terms*, International Telecommunication Union, Geneva.

A suffix "1" after a definition below indicates that the definition is taken from the *Data Communications Glossary*, International Business Machines Corporation, Poughkeepsie, 1967 (Manual number C20–1666).

Address. A coded representation of the destination of data, or of their originating terminal. Multiple terminals on one communication line, for example, must have unique addresses. Telegraph messages reaching a switching center carry an address before their text to indicate the destination of the message.

Alphabet (telegraph or data). A table of correspondence between an agreed set of characters and the signals which represent them. (2).

Alternate routing. An alternative communications path used if the normal one is not available. There may be one or more possible alternative paths.

Amplitude modulation. One of three ways of modifying a sine wave signal in order to make it "carry" information. The sine wave, or "carrier," has its amplitude modified in accordance with the information to be transmitted.

Analog data. Data in the form of *continuously variable* physical quantities. (Compare with **Digital data**.) (1).

Analog transmission. Transmission of a continuously variable signal as opposed to a discretely variable signal. Physical quantities such as temperature are continuously variable and so are described as "analog." Data characters, on the other hand, are coded in discrete separate pulses or signal levels, and are referred to as "digital." The normal way of transmitting a telephone, or voice, signal has been analog; but now digital encoding (using **PCM**) is coming into use over trunks.

381

Application program. The working programs in a system may be classed as *application programs* and *supervisory programs*. The application programs are the main data-processing programs. They contain no input-output coding except in the form of macroinstructions that transfer control to the supervisory programs. They are usually unique to one type of application, whereas the supervisory programs could be used for a variety of different application types. A number of different terms are used for these two classes of program.

ARQ (Automatic Request for Repetition). A system employing an error-detecting code and so conceived that any false signal initiates a repetition of the transmission of the character incorrectly received. (2).

ASCII (American Standard Code for Information Interchange). Usually pronounced "ask'-ee." An eight-level code for data transfer adopted by the American Standards Association to achieve compatibility between data devices. (1).

Asynchronous transmission. Transmission in which each information character, or sometimes each word or small block, is individually synchronized, usually by the use of start and stop elements. The gap between each character (or word) is not of a necessarily fixed length. (Compare with **Synchronous transmission.**) Asynchronous transmission is also called *start-stop transmission*.

Attended operation. In data set applications, individuals are required at both stations to establish the connection and transfer the data sets from talk (voice) mode to data mode. (Compare **Unattended operation.**) (1).

Attenuation. Decrease in magnitude of current, voltage, or power of a signal in transmission between points. May be expressed in decibels. (1).

Atenuation equalizer. (*See* **Equalizer.**)

Audio frequencies. Frequencies that can be heard by the human ear (usually 30 to 20,000 cycles per second). (1).

Automatic calling unit (ACU). A dialing device supplied by the communications common carrier, which permits a business machine to automatically dial calls over the communication networks. (1).

Automatic dialing unit (ADU). A device capable of automatically generating dialing digits. (Compare with **Automatic calling unit.**) (1).

Bandwidth. The range of frequencies available for signaling. The difference expressed in cycles per second (hertz) between the highest and lowest frequencies of a band.

Baseband signaling. Transmission of a signal at its original frequencies, i.e., a signal not changed by modulation.

Baud. Unit of signaling speed. The speed in bauds is the number of discrete conditions or signal events per second. (This is applied only to the actual signals on a communication line.) If each signal event represents only one bit condition, baud is the same as bits per second. When each signal event represents other than one bit (e.g., see **Dibit**), baud does not equal bits per second. (1).

Baudot code. A code for the transmission of data in which five equal-length bits represent one character. This code is used in most DC teletypewriter machines where 1 start element and 1.42 stop elements are added. (See page 109.) (1).

Bel. Ten decibels, q.v.

BEX. Broadband exchange, q.v.

Bias distortion. In teletypewriter applications, the uniform shifting of the beginning of all marking pulses from their proper positions in relation to the beginning of the start pulse. (1).

Bias distortion, asymmetrical distortion. Distortion affecting a two-condition (or binary) modulation (or restitution) in which all the significant conditions have longer or shorter durations than the corresponding theoretical durations. (2).

Bit. Contraction of "binary digit," the smallest unit of information in a binary system. A bit represents the choice between a mark or space (one or zero) condition.

Bit rate. The speed at which bits are transmitted, usually expressed in bits per second. (Compare with **Baud.**)

Broadband. Communication channel having a bandwidth greater than a voice-grade channel, and therefore capable of higher-speed data transmission. (1).

Broadband exchange (BEX). Public switched communication system of Western Union, featuring various bandwidth FDX connections. (1).

Buffer. A storage device used to compensate for a difference in rate of data flow, or time of occurrence of events, when transmitting data from one device to another. (1).

Cable. Assembly of one or more conductors within an enveloping protective sheath, so constructed as to permit the use of conductors separately or in groups. (1).

Carrier. A continuous frequency capable of being modulated, or impressed with a second (information carrying) signal. (1).

Carrier, communications common. A company which furnishes communications services to the general public, and which is regulated by appropriate local, state, or federal agencies. The term strictly includes truckers and movers, bus lines, and airlines, but is usually used to refer to telecommunication companies.

Carrier system. A means of obtaining a number of channels over a single path by modulating each channel on a different carrier frequency and demodulating at the receiving point to restore the signals to their original form.

Carrier telegraphy, carrier current telegraphy. A method of transmission in which the signals from a telegraph transmitter modulate an alternating current. (2).

Central office. The place where communications common carriers terminate customer lines and locate the switching equipment which interconnects those lines. (Also referred to as an *exchange, end office*, and *local central office.*)

Chad. The material removed when forming a hole or notch in a storage medium such as punched tape or punched cards.

Chadless tape. Perforated tape with the chad partially attached, to facilitate interpretive printing on the tape.

Channel. 1. (CCITT and ASA standard) A means of one-way transmission. (Compare with **Circuit**.)
2. (Tariff and common usage) As used in the tariffs, a path for electrical transmission between two or more points without common-carrier-provided terminal equipment. Also called *circuit, line, link, path,* or *facility.* (1).

Channel, analog. A channel on which the information transmitted can take any value between the limits defined by the channel. Most voice channels are analog channels.

Channel, voice-grade. A channel suitable for transmission of speech, digital or analog data, or facsimile, generally with a frequency range of about 300 to 3400 cycles per second.

12-channel group (of carrier current system). The assembly of 12 telephone channels, in a carrier system, occupying adjacent bands in the spectrum, for the purpose of simultaneous modulation or demodulation. (2).

Character. Letter, figure, number, punctuation or other sign contained in a message. Besides such characters, there may be characters for special symbols and some control functions. (1).

Characteristic distortion. Distortion caused by transients which, as a result of the modulation, are present in the transmission channel and depend on its transmission qualities.

Circuit. A means of both-way communication between two points, comprising associated "go" and "return" channels. (1).

Circuit, four-wire. A communication path in which four wires (two for each direction of transmission) are presented to the station equipment. (1).

Circuit, two-wire. A metallic circuit formed by two conductors insulated from each other. It is possible to use the two conductors as either a one-way transmission path, a half-duplex path, or a duplex path. (1).

Common carrier. (*See* **Carrier, communications common**.)

Compandor. A compandor is a combination of a compressor at one point in a communication path for reducing the volume *range* of signals, followed by an expandor at another point for restoring the original volume range. Usually its purpose is to improve the ratio of the signal to the interference entering in the path between the compressor and expandor. (2).

Compressor. Electronic device which compresses the volume range of a signal, used in a compandor (q.v.). An "expandor" restores the original volume range after transmission.

Conditioning. The addition of equipment to a leased voice-grade channel to provide minimum values of line characteristics required for data transmission. (1).

Contention. This is a method of line control in which the terminals request to transmit. If the channel in question is free, transmission goes ahead; if it is not free, the terminal will have to wait until it becomes free. The queue of contention requests may be built up by the computer, and this can either be in a prearranged sequence or in the sequence in which the requests are made.

Control character. A character whose occurrence in a particular context initiates, modifies, or stops a control operation—e.g., a character to control carriage return. (1).

Control mode. The state that all terminals on a line must be in to allow line control actions, or terminal selection to occur. When all terminals on a line are in the control mode, characters on the line are viewed as control characters performing line discipline, that is, polling or addressing. (1).

Cross-bar switch. A switch having a plurality of vertical paths, a plurality of horizontal paths, and electromagnetically operated mechanical means for interconnecting any one of the vertical paths with any of the horizontal paths. See page 326. (2).

Cross-bar system. A type of line-switching system which uses cross-bar switches.

Cross talk. The unwanted transfer of energy from one circuit, called the *disturbing* circuit, to another circuit, called the *disturbed* circuit. (2).

Cross talk, far-end. Cross talk which travels along the disturbed circuit in the same direction as the signals in that circuit. To determine the far-end cross talk between two pairs, 1 and 2, signals are transmitted on pair 1 at station A, and the level of cross talk is measured on pair 2 at station B. (1).

Cross talk, near-end. Cross talk which is propagated in a disturbed channel in the direction opposite to the direction of propagation of the current in the distrubing channel. Ordinarily, the terminal of the disturbed channel at which the near-end cross talk is present is near or coincides with the energized terminal of the disturbing channel. (1).

Dataphone. Both a service mark and a trademark of AT & T and the Bell System. As a service mark it indicates the transmission of data over the telephone network. As a trademark it identifies the communications equipment furnished by the Bell System for data communications services. (1).

Data set. A device which performs the modulation/demodulation and control functions necessary to provide compatibility between business machines and communications facilities. (*See also* **Line adapter, Modem,** *and* **Subset.**) (1).

Data-signaling rate. It is given by $\sum_{i=1}^{m} \frac{1}{T_i} \log_2 n_i$, where m is the number of parallel channels, T is the minimum interval for the ith channel, expressed in seconds, n is the number of significant conditions of the modulation in the ith channel. Data-signaling rate is expressed in bits per second. (2).

Dataspeed. An AT&T marketing term for a family of medium-speed paper tape transmitting and receiving units. Similar equipment is also marketed by Western Union. (1).

DDD. (*See* **Direct distance dialing,** q.v.)

Decibel (db). A tenth of a bel. A unit for measuring relative strength of a signal parameter such as power, voltage, etc. The number of decibels is ten times the logarithm (base 10) of the ratio of the measured quantity to the reference level. The reference level must always be indicated, such as 1 milliwatt for power ratio. (1). See Fig. 9.5.

Delay distortion. Distortion occurring when the envelope delay of a circuit or system is not constant over the frequency range required for transmission.

Delay equalizer. A corrective network which is designed to make the phase delay or envelope delay of a circuit or system substantially constant over a desired frequency range. (*See* **Equalizer.**) (1).

Demodulation. The process of retrieving intelligence (data) from a modulated carrier wave; the reverse of modulation. (1).

Diagnostic programs. These are used to check equipment malfunctions and to pinpoint faulty components. They may be used by the computer engineer or may be called in by the supervisory programs automatically.

Diagnostics, system. Rather than checking one individual component, system diagnostics utilize the whole system in a manner similar to its operational running. Programs resembling the operational programs will be used rather than systematic programs that run logical patterns. These will normally detect overall system malfunctions but will not isolate faulty components.

Diagnostics, unit. These are used on a conventional computer to detect faults in the various units. Separate unit diagnostics will check such items as arithmetic circuitry, transfer instructions, each input-output unit, and so on.

Dial pulse. A current interruption in the DC loop of a calling telephone. It is produced by the breaking and making of the dial pulse contacts of a calling telephone when a digit is dialed. The loop current is interrupted once for each unit of value of the digit. (1).

Dial-up. The use of a dial or pushbutton telephone to initiate a station-to-station telephone call.

Dibit. A group of two bits. In four-phase modulation, each possible dibit is encoded as one of four unique carrier phase shifts. The four possible states for a dibit are 00, 01, 10, 11.

Differential modulations. A type of modulation in which the choice of the significant condition for any signal element is dependent on the choice for the previous signal element. (2).

Digital data. Information represented by a code consisting of a sequence of discrete elements. (Compare with **Analog data.**) (1).

Digital signal. A discrete or discontinuous signal; one whose various states are discrete intervals apart. (Compare with **Analog transmission.**) (1).

Direct distance dialing (DDD). A telephone exchange service which enables the telephone user to call other subscribers outside his local area without operator assistance. In the United Kingdom and some other countries, this is called *Subscriber Trunk Dialing* (STD).

Disconnect signal. A signal transmitted from one end of a subscriber line or trunk to indicate at the other end that the established connection should be disconnected. (1).

Distortion. The unwanted change in waveform that occurs between two points in a transmission system. (1).

Distributing frame. A structure for terminating permanent wires of a telephone central office, private branch exchange, or private exchange, and for permitting the easy change of connections between them by means of cross-connecting wires. (1).

Double-current transmission, polar direct-current system. A form of binary telegraph transmission in which positive and negative direct currents denote the significant conditions. (2).

Drop, subscriber's. The line from a telephone cable to a subscriber's building. (1).

Duplex transmission. Simultaneous two-way independent transmission in both directions. (Compare with **Half-duplex transmission.** Also called *full-duplex transmission.*) (1).

Duplexing. The use of duplicate computers, files or circuitry, so that in the event of one component failing an alternative one can enable the system to carry on its work.

Echo. An echo is a wave which has been reflected or otherwise returned with sufficient magnitude and delay for it to be perceptible in some manner as a wave distinct from that directly transmitted.

Echo check. A method of checking data transmission accuracy whereby the received data are returned to the sending end for comparison with the original data.

Echo suppressor. A line device used to prevent energy from being reflected back (echoed) to the transmitter. It attenuates the transmission path in one direction while signals are being passed in the other direction. (1).

End distortion. End distortion of start-stop teletypewriter signals is the shifting of the end of all marking pulses from their proper positions in relation to the beginning of the start pulse.

End office. (*See* **Central office.**)

Equalization. Compensation for the attenuation (signal loss) increase with frequency. Its purpose is to produce a flat frequency response while the temperature remains constant. (1).

Equalizer. Any combination (usually adjustable) of coils, capacitors, and/or resistors inserted in transmission line or amplifier circuit to improve its frequency response. (1).

Equivalent four-wire system. A transmission system using frequency division to obtain full-duplex operation over only one pair of wires. (1).

Error-correcting telegraph code. An error-detecting code incorporating sufficient additional signaling elements to enable the nature of some or all of the errors to be indicated and corrected entirely at the receiving end.

Error-detecting and feedback system, decision feedback system, request repeat system, ARQ system. A system employing an error-detecting code and so arranged that a signal detected as being in error automatically initiates a request for retransmission of the signal detected as being in error. (2).

Error-detecting telegraph code. A telegraph code in which each telegraph signal conforms to specific rules of construction, so that departures from this construction in the received signals can be automatically detected. Such codes necessarily require more signaling elements than are required to convey the basic information.

ESS. (Electronic Switching System). Bell System term for computerized telephone exchange. ESS 1 is a central office. ESS 101 gives private branch exchange (PBX) switching controlled from the local central office. (*See* Chapter 19.)

Even parity check (odd parity check). This is a check which tests whether the number of digits in a group of binary digits is even (even parity check) or odd (odd parity check). (2).

Exchange. A unit established by a communications common carrier for the administration of communication service in a specified area which usually embraces a city, town, or village and its environs. It consists of one or more central offices together with the associated equipment used in furnishing communication service. (This term is often used as a synonym for "central office," q.v.)

Exchange, classes of. Class 1 (*see* **Regional center**); class 2 (*see* **Sectional center**); class 3 (*see* **Primary center**); class 4 (*see* **Toll center**); class 5 (*see* **End office**).

Exchange, private automatic (PAX). A dial telephone exchange that provides private telephone service to an organization and that does *not* allow calls to be transmitted to or from the public telephone network.

Exchange, private automatic branch (PABX). A private automatic telephone exchange that provides for the transmission of calls to and from the public telephone network.

Exchange, private branch (PBX). A manual exchange connected to the public telephone network on the user's premises and operated by an attendant supplied by the user. PBX is today commonly used to refer also to an automatic exchange.

Exchange, trunk. An exchange devoted primarily to interconnecting trunks.

Exchange service. A service permitting interconnection of any two customers' stations through the use of the exchange system.

Expandor. A transducer which for a given amplitude range of input voltages produces a larger range of output voltages. One important type of expandor employs the information from the envelope of speech signals to expand their volume range. (Compare **Compandor.**) (1).

Facsimile (FAX). A system for the transmission of images. The image is scanned at the transmitter, reconstructed at the receiving station, and duplicated on some form of paper. (1).

Fail softly. When a piece of equipment fails, the programs let the system fall back to a degraded mode of operation rather than let it fail catastrophically and give no response to its users.

Fall-back, double. Fall-back in which two separate equipment failures have to be contended with.

Fall-back procedures. When the equipment develops a fault the programs operate in such a way as to circumvent this fault. This may or may not give a degraded service. Procedures necessary for fall-back may include those to switch over to an alternative computer or file, to change file addresses, to send output to a typewriter instead of a printer, to use different communication lines or bypass a faulty terminal, etc.

FCC. Federal Communications Commission, q.v.

FD or **FDX.** Full duplex. (*See* **Duplex.**)

FDM. Frequency-division multiplex, q.v.

Federal Communications Commission (FCC). A board of seven commissioners appointed by the President under the Communication Act of 1934, having the power to regulate all interstate and foreign electrical communication systems originating in the United States. (1).

Figures shift. A physical shift in a teletypewriter which enables the printing of numbers, symbols, upper-case characters, etc. (Compare with **Letters shift.**) (1).

Filter. A network designed to transmit currents of frequencies within one or more frequency bands and to attenuate currents of other frequencies. (2).

Foreign exchange service. A service which connects a customer's telephone to a telephone company central office normally not serving the customer's location. (Also applies to TWX service.) (1).

Fortuitous distortion. Distortion resulting from causes generally subject to random laws (accidental irregularities in the operation of the apparatus and of the moving parts, disturbances affecting the transmission channel, etc.). (2).

Four-wire circuit. A circuit using two pairs of conductors, one pair for the "go" channel and the other pair for the "return" channel. (2).

Four-wire equivalent circuit. A circuit using the same pair of conductors to give "go" and "return" channels by means of different carrier frequencies for the two channels. (2).

Four-wire terminating set. Hybrid arrangement by which four-wire circuits are terminated on a two-wire basis for interconnection with two-wire circuits.

Frequency-derived channel. Any of the channels obtained from multiplexing a channel by frequency division. (2).

Frequency-division multiplex. A multiplex system in which the available transmission frequency range is divided into narrower bands, each used for a separate channel. (2).

Frequency modulation. One of three ways of modifying a sine wave signal to make it "carry" information. The sine wave or "carrier" has its frequency modified in accordance with the information to be transmitted. The frequency function of the modulated wave may be continuous or discontinuous. In the latter case, two or more particular frequencies may correspond each to one significant condition.

Frequency-shift signaling, frequency-shift keying (FSK). Frequency modulation method in which the frequency is made to vary at the significant instants. 1. By smooth transitions: the modulated wave and the change in frequency are continuous at the significant instants. 2. By abrupt transitions: the modulated wave is continuous but the frequency is discontinuous at the significant instants. (2).

FSK. Frequency-shift keying, q.v.

FTS. Federal Telecommunications System.

Full-duplex (FD or FDX) **transmission.** (*See* **Duplex transmission.**)

Half-duplex (HD or HDX) **circuit.**
1. CCITT definition: A circuit designed for duplex operation, but which, on account of the nature of the terminal equipments, can be operated alternately only.
2. Definition in common usage (the normal meaning in computer literature): A circuit designed for transmission in either direction but not both directions simultaneously.

Handshaking. Exchange of predetermined signals for purposes of control when a connection is established between two data sets.

Harmonic distortion. The resultant presence of harmonic frequencies (due to non-linear characteristics of a transmission line) in the response when a sinusoidal stimulus is applied. (1)

HD or HDX. Half duplex. (*See* **Half-duplex circuit.**)

Hertz (Hz). A measure of frequency or bandwidth. The same as cycles per second.

Home loop. An operation involving only those input and output units associated with the local terminal. (1).

In-house. *See* **In-plant system.**

In-plant system. A system whose parts, including remote terminals, are all situated in one building or localized area. The term is also used for communication systems spanning several buildings and sometimes covering a large distance, but in which no common carrier facilities are used.

International Telecommunication Union (ITU). The telecommunications agency of the United Nations, established to provide standardized communications procedures and practices including frequency allocation and radio regulations on a world-wide basis.

Interoffice trunk. A direct trunk between local central offices.

Intertoll trunk. A trunk between toll offices in different telephone exchanges. (1).

ITU. International Telecommunication Union, q.v.

Keyboard perforator. A perforator provided with a bank of keys, the manual depression of any one of which will cause the code of the corresponding character or function to be punched in a tape. (2).

Keyboard send/receive. A combination teletypewriter transmitter and receiver with transmission capability from keyboard only.

KSR. Keyboard send/receive, q.v.

Leased facility. A facility reserved for sole use of a single leasing customer. (*See also* **private line.**) (1).

Letters shift. A physical shift in a teletypewriter which enables the printing of alphabetic characters. Also, the name of the character which causes this shift. (*Compare* with **Figures shift.**) (1).

Line switching. Switching in which a circuit path is set up between the incoming and outgoing lines. Contrast with message switching (q.v.) in which no such physical path is established.

Link communication. The physical means of connecting one location to another for the purpose of transmitting and receiving information. (1).

Loading. Adding inductance (load coils) to a transmission line to minimize amplitude distortion. (1).

Local exchange, local central office. An exchange in which subscribers' lines terminate. (Also referred to as *end office.*)

Local line, local loop. A channel connecting the subscriber's equipment to the line terminating equipment in the central office exchange. Usually metallic circuit (either two-wire or four-wire). (1).

Longitudinal redundancy check (LRC). A system of error control based on the formation of a block check following preset rules. The check formation rule is applied in the same manner to each character. In a simple case, the LRC is created by forming a parity check on each bit position of all the characters in the block (e.g., the first bit of the LRC character creates odd parity among the one-bit positions of the characters in the block).

Loop checking, message feedback, information feedback. A method of checking the accuracy of transmission of data in which the received data are returned to the sending end for comparison with the original data, which are stored there for this purpose. (2).

LRC. Longitudinal redundancy check.

LTRS. Letters shift, q.v. (*See* **Letters shift.**)

Mark. Presence of signal. In telegraph communications a mark represents the closed condition or current flowing. A mark impulse is equivalent to a binary 1.

Mark-hold. The normal no-traffic line condition whereby a steady mark is transmitted.

Mark-to-space transition. The transition, or switching from a marking impulse to a spacing impulse.

Mark-hold. The normal no-traffic line condition whereby a steady mark is transmitted. This may be a customer-selectable option. (Compare with **Space-hold.**) (1).

Master station. A unit having control of all other terminals on a multipoint circuit for purposes of polling and/or selection. (1).

Mean time to failure. The average length of time for which the system, or a component of the system, works without fault.

Mean time to repair. When the system, or a component of the system, develops a fault, this is the average time taken to correct the fault.

Message reference block. When more than one message in the system is being processed in parallel, an area of storage is allocated to each message and remains uniquely associated with that message for the duration of its stay in the computer. This is called the *message reference block* in this book. It will normally contain the message and data associated with it that are required for its processing. In most systems, it contains an area of working storage uniquely reserved for that message.

Message switching. The technique of receiving a message, storing it until the proper outgoing line is available, and then retransmitting. No direct connection between the incoming and outgoing lines is set up as in line switching (q.v.).

Microwave. Any electromagnetic wave in the radio-frequency spectrum above 890 megacycles per second. (1).

Modem. A contraction of "modulator-demodulator." The term may be used when the modulator and the demodulator are associated in the same signal-conversion equipment. (*See* **Modulation** *and* **Data set.**) (1).

Modulation. The process by which some characteristic of one wave is varied in accordance with another wave or signal. This technique is used in data sets and modems to make business machine signals compatible with communications facilities. (1).

Modulation with a fixed reference. A type of modulation in which the choice of the significant condition for any signal element is based on a fixed reference. (2).

Multidrop line. Line or circuit interconnecting several stations. (Also called *multipoint line.*) (1).

Multiplex, multichannel. Use of a common channel in order to make two or more channels, either by splitting of the frequency band transmitted by the common channel into narrower bands, each of which is used to constitute a distinct channel (frequency-division multiplex), or by allotting this common channel in turn, to constitute different intermittent channels (time-division multiplex). (2).

Multiplexing. The division of a transmission facility into two or more channels either by splitting the frequency band transmitted by the channel into narrower bands, each of which is used to constitute a distinct channel (frequency-division multiplex), or by allotting this common channel to several different information channels, one at a time (time-division multiplexing). (2).

Multiplexor. A device which uses several communication channels at the same time, and transmits and receives messages and controls the communication lines. This device itself may or may not be a stored-program computer.

Multipoint line. (*See* **Multidrop line.**)

Neutral transmission. Method of transmitting teletypewriter signals, whereby a mark is represented by current on the line and a space is represented by the absence of current. By extension to tone signaling, neutral transmission is a method of signaling employing two signaling states, one of the states representing both a space condition and also the absence of any signaling. (Also called *unipolar.* Compare with **Polar transmission.**) (1).

Noise. Random electrical signals, introduced by circuit components or natural disturbances, which tend to degrade the performance of a communications channel. (1).

Off hook. Activated (in regard to a telephone set). By extension, a data set automatically answering on a public switched system is said to go "off hook." (Compare with **On hook.**) (1).

Off line. Not in the line loop. In telegraph usage, paper tapes frequently are punched "off line" and then transmitted using a paper tape transmitter.

On hook. Deactivated (in regard to a telephone set). A telephone not in use is "on hook." (1).

On line. Directly in the line loop. In telegraph usage, transmitting directly onto the line rather than, for example, perforating a tape for later transmission. (*See also* **On-line computer system.**)

On-line computer system. An on-line system may be defined as one in which the input data enter the computer directly from their point of origin and/or output data are transmitted directly to where they are used. The intermediate stages such as punching data into cards or paper tape, writing magnetic tape, or off-line printing, are largely avoided.

Open wire. A conductor separately supported above the surface of the ground— i.e., supported on insulators.

Open-wire line. A pole line whose conductors are principally in the form of open wire.

PABX. Private automatic branch exchange. (*See* **Exchange, private automatic branch.**)

Parallel transmission. Simultaneous transmission of the bits making up a character or byte, either over separate channels or on different carrier frequencies on the channel. (1). The simultaneous transmission of a certain number of signal elements constituting the same telegraph or data signal. For example, use of a code according to which each signal is characterized by a combination of 3 out of 12 frequencies simultaneously transmitted over the channel. (2).

Parity check. Addition of noninformation bits to data, making the number of ones in a grouping of bits either always even or always odd. This permits detection of bit groupings that contain single errors. It may be applied to characters, blocks, or any convenient bit grouping. (1).

Parity check, horizontal. A parity check applied to the group of certain bits from every character in a block. (*See also* **Longitudinal redundancy check.**)

Parity check, vertical. A parity check applied to the group which is all bits in one character. (Also called *vertical redundancy check.*) (1).

PAX. Private automatic exchange. (*See* **Exchange, private automatic.**)

PBX. Private branch exchange. (*See* **Exchange, private branch.**)

PCM. (*See* **Pulse code modulation.**)

PDM. (*See* **Pulse duration modulation.**)

Perforator. An instrument for the manual preparation of a perforated tape, in which telegraph signals are represented by holes punched in accordance with a predetermined code. Paper tape is prepared off line with this. (Compare with **Reperforator.**) (2).

Phantom telegraph circuit. Telegraph circuit superimposed on two physical circuits reserved for telephony. (2).

Phase distortion. (*See* **Distortion, delay.**)

Phase equalizer, delay equalizer. A delay equalizer is a corrective network which is designed to make the phase delay or envelope delay of a circuit or system substantially constant over a desired frequency range. (2).

Phase-inversion modulation. A method of phase modulation in which the two significant conditions differ in phase by π radians. (2).

Phase modulation. One of three ways of modifying a sine wave signal to make it "carry" information. The sine wave or "carrier," has its phase changed in accordance with the information to be transmitted.

Pilot model. This is a model of the system used for program testing purposes which is less complex than the complete model, e.g., the files used on a pilot model may contain a much smaller number of records than the operational files; there may be few lines and fewer terminals per line.

Polar transmission. A method for transmitting teletypewriter signals, whereby the marking signal is represented by direct current flowing in one direction and the spacing signal is represented by an equal current flowing in the opposite direction. By extension to tone signaling, polar transmission is a method of transmission employing three distinct states, two to represent a mark and a space and one to represent the absence of a signal. (Also called *bipolar.* Compare with **Neutral transmission.**)

Polling. This is a means of controlling communication lines. The communication control device will send signals to a terminal saying, "Terminal A. Have you anything to send?" if not, "Terminal B. Have you anything to send?" and so on. Polling is an alternative to contention. It makes sure that no terminal is kept waiting for a long time.

Polling list. The polling signal will usually be sent under program control. The program will have in core a list for each channel which tells the sequence in which the terminals are to be polled.

PPM. (*See* **Pulse position modulation.**)

Primary center. A control center connecting toll centers; a class 3 office. It can also serve as a toll center for its local end offices.

Private automatic branch exchange. (*See* **Exchange, private automatic branch.**)

Private automatic exchange. (*See* **Exchange, private automatic.**)

Private branch exchange (PBX). A telephone exchange serving an individual organization and having connections to a public telephone exchange. (2).

Private line. Denotes the channel and channel equipment furnished to a customer as a unit for his exclusive use, without interexchange switching arrangements. (1).

Processing, batch. A method of computer operation in which a number of similar input items are accumulated and grouped for processing.

Processing, in line. The processing of transactions as they occur, with no preliminary editing or sorting of them before they enter the system. (1).

Propagation delay. The time necessary for a signal to travel from one point on a circuit to another.

Public. Provided by a common carrier for use by many customers.

Public switched network. Any switching system that provides circuit switching to many customers. In the U.S.A. there are four such networks: Telex, TWX, telephone, and Broadband Exchange. (1).

Pulse-code modulation (PCM). Modulation of a pulse train in accordance with a code. (2).

Pulse-duration modulation (PDM) (**pulse-width modulation**) (**pulse-length modulation**). A form of pulse modulation in which the durations of pulses are varied. (2).

Pulse modulation. Transmission of information by modulation of a pulsed or intermittent, carrier. Pulse width, count, position, phase, and/or amplitude may be the varied characteristic.

Pulse-position modulation (PPM). A form of pulse modulation in which the positions in time of pulses are varied, without modifying their duration. (2).

Pushbutton dialing. The use of keys or pushbuttons instead of a rotary dial to generate a sequence of digits to establish a circuit connection. The signal form is usually multiple tones. (Also called *tone dialing, Touch-call, Touch-Tone.*) (1).

Real time. A real-time computer system may be defined as one that controls an environment by receiving data, processing them, and returning the results sufficiently quickly to affect the functioning of the environment at that time.

Reasonableness checks. Tests made on information reaching a real-time system or being transmitted from it to ensure that the data in question lie within a given range. It is one of the means of protecting a system from data transmission errors.

Recovery from fall-back. When the system has switched to a fall-back mode of operation and the cause of the fall-back has been removed, the system must be restored to its former condition. This is referred to as *recovery from fall-back.* The recovery process may involve updating information in the files to produce two duplicate copies of the file.

Redundancy check. An automatic or programmed check based on the systematic insertion of components or characters used especially for checking purposes. (1).

Redundant code. A code using more signal elements than necessary to represent the intrinsic information. For example, five-unit code using all the characters of International Telegraph Alphabet No. 2 is not redundant; five-unit code using only the figures in International Telegraph Alphabet No. 2 is redundant; seven-unit code using only signals made of four "space" and three "mark" elements is redundant. (2).

Reference pilot. A reference pilot is a different wave from those which transmit the telecommunication signals (telegraphy, telephony). It is used in carrier systems to facilitate the maintenance and adjustment of the carrier transmission system. (For example, automatic level regualtion, synchronization of oscillators, etc.) (2).

Regenerative repeater. (*See* **Repeater, regenerative.**)

Regional center. A control center (class 1 office) connecting sectional centers of the telephone system together. Every pair of regional centers in the United States has a direct circuit group running from one center to the other. (1).

Repeater.
1. A device whereby currents received over one circuit are automatically repeated in another circuit or circuits, generally in an amplified and/or reshaped form.
2. A device used to restore signals, which have been distorted because of attenuation, to their original shape and transmission level.

Repeater, regenerative. Normally, a repeater utilized in telegraph applications. Its function is to retime and retransmit the received signal impulses restored to their original strength. These repeaters are speed- and code-sensitive and are intended for use with standard telegraph speeds and codes. (Also called *regen.*) (1).

Repeater, telegraph. A device which receives telegraph signals and automatically retransmits corresponding signals. (2).

Reperforator (receiving perforator). A telegraph instrument in which the received signals cause the code of the corresponding characters or functions to be punched in a tape. (1).

Reperforator/transmitter (RT). A teletypewriter unit consisting of a reperforator and a tape transmitter, each independent of the other. It is used as a relaying device and is especially suitable for transforming the incoming speed to a different outgoing speed, and for temporary queuing.

Residual error rate, undetected error rate. The ratio of the number of bits, unit elements, characters or blocks incorrectly received but undetected or uncorrected by the error-control equipment, to the total number of bits, unit elements, characters or blocks sent. (2).

Response time. This is the time the system takes to react to a given input. If a message is keyed into a terminal by an operator and the reply from the computer, when it comes, is typed at the same terminal, response time may be defined as the time interval between the operator pressing the last key and the terminal typing the first letter of the reply. For different types of terminal, response time may be defined similarly. It is the interval between an event and the system's response to the event.

Ringdown. A method of signaling subscribers and operators using either a 20-cycle AC signal, a 135-cycle AC signal, or a 1000-cycle signal interrupted 20 times per second. (1).

Routing. The assignment of the communications path by which a message or telephone call will reach its destination. (1).

Routing, alternate. Assignment of a secondary communications path to a destination when the primary path is unavailable. (1).

Routing indicator. An address, or group of characters, in the heading of a message defining the final circuit or terminal to which the message has to be delivered. (1).

RT. Reperforator/transmitter, q.v.

Saturation testing. Program testing with a large bulk of messages intended to bring to light those errors which will only occur very infrequently and which may be triggered by rare coincidences such as two different messages arriving at the same time.

Sectional center. A control center connecting primary centers; a class 2 office. (1).

Seek. A mechanical movement involved in locating a record in a random-access file. This may, for example, be the movement of an arm and head mechanism that is necessary before a read instruction can be given to read data in a certain location on the file.

Selection. Addressing a terminal and/or a component on a selective calling circuit. (1).

Selective calling. The ability of the transmitting station to specify which of several stations on the same line is to receive a message. (1).

Self-checking numbers. Numbers which contain redundant information so that an error in them, caused, for example, by noise on a transmission line, may be detected.

Serial transmission. Used to identify a system wherein the bits of a character occur serially in time. Implies only a single transmission channel. (Also called *serial-by-bit*.) (1). Transmission at successive intervals of signal elements constituting the same telegraph or data signal. For example, transmission of signal elements by a standard teleprinter, in accordance with International Telegraph Alphabet No. 2; telegraph transmission by a time-divided channel. (2).

Sideband. The frequency band on either the upper or lower side of the carrier frequency within which fall the frequencies produced by the process of modulation. (2).

Signal-to-noise ratio (S/N). Relative power of the signal to the noise in a channel. (1).

Simplex circuit.
1. CCITT definition: A circuit permitting the transmission of signals in either direction, but not in both simultaneously.
2. Definition in common usage (the normal meaning in computer literature): A circuit permitting transmission in one specific direction only.

Simplex mode. Operation of a communication channel in one direction only, with no capability for reversing. (1).

Simulation. This is a word which is sometimes confusing as it has three entirely different meanings, namely:

Simulation for design and monitoring. This is a technique whereby a model of the working system can be built in the form of a computer program. Special computer languages are available for producing this model. A complete system may be described by a succession of different models. These models can then be adjusted easily and endlessly, and the system that is being designed or monitored can be experimented with to test the effect of any proposed changes. The simulation model is a program that is run on a computer separate from the system that is being designed.

Simulation of input devices. This is a program testing aid. For various reasons it is undesirable to use actual lines and terminals for some of the program testing. Therefore, magnetic tape or other media may be used and read in by a special program which makes the data appear as if they came from actual lines and terminals. Simulation in this sense is the replacement of one set of equipment by another set of equipment and programs, so that the behavior is similar.

Simulation of superivsory programs. This is used for program testing purposes when the actual supervisory programs are not yet available. A comparatively simple program to bridge the gap is used instead. This type of simulation is the replacement of one set of programs by another set which imitates it.

Single-current transmission, (inverse) **neutral direct-current system.** A form of telegraph transmission effected by means of unidirectional currents. (2).

Space. 1. An impulse which, in a neutral circuit, causes the loop to open or causes absence of signal, while in a polar circuit it causes the loop current to flow in a direction opposite to that for a mark impulse. A space impulse is equivalent to a binary 0. 2. In some codes, a character which causes a printer to leave a character width with no printed symbol. (1).

Space-hold. The normal no-traffic line condition whereby a steady space is transmitted. (Compare with **Mark-hold.**) (1).

Space-to-mark transition. The transition, or switching, from a spacing impulse to a marking impulse. (1).

Spacing bias. *See* **Distortion, bias.**

Spectrum. 1. A continuous range of frequencies, usually wide in extent, within which waves have some specific common characteristic. 2. A graphical representation of the distribution of the amplitude (and sometimes phase) of the components of a wave as a function of frequency. A spectrum may be continuous or, on the contrary, contain only points corresponding to certain discrete values. (2).

Start element. The first element of a character in certain serial transmissions, used to permit synchronization, In Baudot teletypewriter operation, it is one space bit. (1).

Start-stop system. A system in which each group of code elements corresponding to an alphabetical signal is preceded by a start signal which serves to prepare the receiving mechanism for the reception and registration of a character, and is followed by a stop signal which serves to bring the receiving mechanism to rest in preparation for the reception of the next character. (Contrast with **Synchronous system.**) (Start-stop transmission is also referred to as *asynchronous transmission*, q.v.)

Station. One of the input or output points of a communications system—e.g., the telephone set in the telephone system or the point where the business machine interfaces the channel on a leased private line. (1).

Status maps. Tables which give the status of various programs, devices, input-output operations, or the status of the communication lines.

Step-by-step switch. A switch that moves in synchronism with a pulse device such as a rotary telephone dial. Each digit dialed causes the movement of successive selector switches to carry the connection forward until the desired line is reached. (Also called *stepper switch*. Compare with **Line switching** and **Cross-bar system.**) (1).

Step-by-step system. A type of line-switching system which uses step-by-step switches. (1).

Stop bit. (*See* **Stop element.**)

Stop element. The last element of a character in asychronous serial transmissions, used to ensure recognition of the next start element. In Baudot teletypewriter operation it is 1.42 mark bits. (*See also* **Start-stop transmission.**) (1.)

Store and forward. The interruption of data flow from the originating terminal to the designated receiver by storing the information enroute and forwarding it at a later time. (*See* **Message switching.**)

Stunt box. A device to 1. control the nonprinting functions of a teletypewriter terminal, such as carriage return and line feed; and 2. a device to recognize line control characters (e.g., DCC, TSC, etc.). (1).

Subscriber trunk dialing. (*See* **direct distance dialing.**)

Subscriber's line. The telephone line connecting the exchange to the subscriber's station. (2).

Subscriber's loop. (*See* **Local loop.**)

Subset. A subscriber set of equipment, such as a telephone. A modulation and demodulation device. (Also called *data set*, which is a more precise term.) (1).

Subscriber's loop. (*See* **Local loop.**)

Subvoice-grade channel. A channel of bandwidth narrower than that of voice-grade channels. Such channels are usually subchannels of a voice-grade line. (1).

Supergroup. The assembly of five 12-channel groups, occupying adjacent bands in the spectrum, for the purpose of simultaneous modulation or demodulation. (2).

Supervisory programs. Those computer programs designed to coordinate service and augment the machine components of the system, and coordinate and service application programs. They handle work scheduling, input-output operations, error actions, and other functions.

Supervisory signals. Signals used to indicate the various operating states of circuit combinations. (1).

Supervisory system. The complete set of supervisory programs used on a given system.

Support programs. The ultimate operational system consists of supervisory programs and application programs. However, a third set of programs are needed to install the system, including diagnostics, testing aids, data generator programs, terminal simulators, etc. These are referred to as *support programs*.

Suppressed carrier transmission. That method of communication in which the carrier frequency is suppressed either partially or to the maximum degree possible. One or both of the sidebands may be transmitted. (1).

Switch hook. A switch on a telephone set, associated with the structure supporting the receiver or handset. It is operated by the removal or replacement of the receiver or handset on the support. (*See also* **Off hook** *and* **On hook.**) (1).

Switching center. A location which terminates multiple circuits and is capable of interconnecting circuits or transferring traffic between circuits; may be automatic, semiautomatic, or torn-tape. (The latter is a location where operators tear off the incoming printed and punched paper tape and transfer it manually to the proper outgoing circuit.)(1).

Switching message. (*See* **Message switching.**)

Switchover. When a failure occurs in the equipment a switch may occur to an alternative component. This may be, for example, an alternative file unit, an alternative communication line or an alternative computer. The switchover process may be automatic under program control or it may be manual.

Synchronous. Having a constant time interval between successive bits, characters, or vents. The term implies that all equipment in the system is in step.

Synchronous system. A system in which the sending and receiving instruments are operating continuously at substantially the same frequency and are maintained, by means of correction, if necessary, in a desired phase relationship. (Contrast with **Start-stop system.**) (2).

Synchronous transmission. A transmission process such that between any two significant instants there is always an integral number of unit intervals. (Contrast with **Asynchronous** or **Start-stop transmission.**) (1).

Tandem office. An office that is used to interconnect the local end offices over tandem trunks in a densely settled exchange area where it is uneconomical for a telephone company to provide direct interconnection between all end offices. The tandem office completes all calls between the end offices but is not directly connected to subscribers. (1).

Tandem office, tandem central office. A central office used primarily as a switching point for traffic between other central offices. (2).

Tariff. The published rate for a specific unit of equipment, facility, or type of service provided by a communications common carrier. Also the vehicle by which the regulating agencies approve or disapprove such facilities or services. Thus the tariff becomes a contract between customer and common carrier.

TD. Transmitter-distributor, q.v.

Teleprocessing. A form of information handling in which a data-processing system utilizes communication facilities. (Originally, but no longer, an IBM trademark.) (1).

Teletype. Trademark of Teletype Corporation, usually referring to a series of different types of teleprinter equipment such as tape punches, reperforators, page printers, etc., utilized for communications systems.

Teletypewriter exchange service (TWX). An AT&T public switched teletypewriter service in which suitably arranged teletypewriter stations are provided with lines to a central office for access to other such stations throughout the U.S.A. and Canada. Both Baudot- and ASCII-coded machines are used. Business machines may also be used, with certain restrictions. (1).

Telex service. A dial-up telegraph service enabling its subscribers to communicate directly and temporarily among themselves by means of start-stop apparatus and of circuits of the public telegraph network. The service operates world wide. Baudot equipment is used. Computers can be connected to the Telex network.

Terminal. Any device capable of sending and/or receiving information over a communication channel. The means by which data are entered into a computer system and by which the decisions of the system are communicated to the environment it affects. A wide variety of terminal devices have been built, including teleprinters, special keyboards, light displays, cathode tubes, thermocouples, pressure gauges and other instrumentation, radar units, telephones, etc.

TEX. (*See* **Telex service.**)

Tie line. A private-line communications channel of the type provided by communications common carriers for linking two or more points together.

Time-derived channel. Any of the channels obtained from multiplexing a channel by time division.

Time-division multiplex. A system in which a channel is established in connecting intermittently, generally at regular intervals and by means of an automatic distribution, its terminal equipment to a common channel. At times when these connections are not established, the section of the common channel between the distributors can be utilized in order to establish other similar channels, in turn.

Toll center. Basic toll switching entity; a central office where channels and toll message circuits terminate. While this is usually one particular central office in a city, larger cities may have several central offices where toll message circuits terminate. A class 4 office. (Also called "toll office" and "toll point.") (1).

Toll circuit (American). *See* **Trunk circuit** (British).

Toll switching trunk (American). *See* **Trunk junction** (British).

Tone dialing. (*See* **Pushbutton dialing.**)

Touch-call. Proprietary term of GT&E. (*See* **Pushbutton dialing.**)

Touch-tone. AT&T term for pushbutton dialing, q.v.

Transceiver. A terminal that can transmit and receive traffic.

Translator. A device that converts information from one system of representation into equivalent information in another system of representation. In telephone equipment, it is the device that converts dialed digits into call-routing information. (1).

Transmitter-distributor (TD). The device in a teletypewriter terminal which makes and breaks the line in timed sequence. Modern usage of the term refers to a paper tape transmitter.

Transreceiver. A terminal that can transmit and receive traffic. (1).

Trunk circuit (British), **toll circuit** (American). A circuit connecting two exchanges in different localities. *Note*: In Great Britain, a trunk circuit is approximately 15 miles long or more. A circuit connecting two exchanges less than 15 miles apart is called a *junction circuit*.

Trunk exchange (British), **toll office** (American). An exchange with the function of controlling the switching of trunk (British) [toll (American)] traffic.

Trunk group. Those trunks between two points both of which are switching centers and/or individual message distribution points, and which employ the same multiplex terminal equipment.

Trunk junction (British), **toll switching trunk** (American). A line connecting a trunk exchange to a local exchange and permitting a trunk operator to call a subscriber to establish a trunk call.

Unattended operations. The automatic features of a station's operation permit the transmission and reception of messages on an unattended basis. (1).

Vertical parity (redundancy) check. (*See* **Parity check, vertical.**)

VOGAD (Voice-Operated Gain-Adjusting Device). A device somewhat similar to a compandor and used on some radio systems; a voice-operated device which removes fluctuation from input speech and sends it out at a constant level. No restoring device is needed at the receiving end. (1).

Voice-frequency, telephone-frequency. Any frequency within that part of the audio-frequency range essential for the transmission of speech of commerical quality, i.e., 300–3400 c/s. (2).

Voice-frequency carrier telegraphy. That form of carrier telegraphy in which the carrier currents have frequencies such that the modulated currents may be transmitted over a voice-frequency telephone channel. (1).

Voice-frequency multichannel telegraphy. Telegraphy using two or more carrier currents the frequencies of which are within the voice-frequency range. Voice-frequency telegraph systems permit the transmission of up to 24 channels over a single circuit by use of frequency-division multiplexing.

Voice-grade channel. (*See* **Channel, voice-grade.**)

Voice-operated device. A device used on a telephone circuit to permit the presence of telephone currents to effect a desired control. Such a device is used in most echo suppressors. (1).

VRC. Vertical redundancey check. (*See also* **Parity check.**)

Watchdog timer. This is a timer which is set by the program. It interrupts the program after a given period of time, e.g., one second. This will prevent the system from going into an endless loop due to a program error, or becoming idle because of an equipment fault. The Watchdog timer may sound a horn or cause a computer interrupt if such a fault is detected.

WATS (Wide Area Telephone Service). A service provided by telephone companies in the United States which permits a customer by use of an access line to make calls to telephones in a specific zone in a dial basis for a flat monthly charge. Monthly charges are based on the size of the area in which the calls are placed, not on the number or length of calls. Under the WATS arrangement, the U.S. is divided into six zones to be called on a full-time or measured-time basis. (1).

Word. 1. In telegraphy, six operations or characters (five characters plus one space). ("Group" is also used in place of "word.") 2. In computing, a sequence of bits or characters treated as a unit and capable of being stored in one computer location. (1).

WPM (Words per minute). A common measure of speed in telegraph systems.

INDEX